CYCLES OF INVENTION AND DISCOVERY

CYCLES OF
INVENTION AND DISCOVERY

. . . .

Rethinking the Endless Frontier

Venkatesh Narayanamurti AND Toluwalogo Odumosu

Harvard University Press

Cambridge, Massachusetts
London, England
2016

Second printing

Library of Congress Cataloging-in-Publication Data
Names: Narayanamurti, Venkatesh, 1939– author. | Odumosu, Toluwalogo,
    1979– author.
Title: Cycles of invention and discovery : rethinking the endless frontier / Venkatesh
    Narayanamurti and Toluwalogo Odumosu.
Description: Cambridge, Massachusetts : Harvard University Press, 2016. | Includes
    bibliographical references and index.
Identifiers: LCCN 2016021048 | ISBN 9780674967960
Subjects: LCSH: Research–Methodology. | Science–Methodology. | Technology–Research.
Classification: LCC Q180.55.M4 N27 2016 | DDC 507.2–dc23
LC record available at https://lccn.loc.gov/2016021048

CONTENTS

1

BREAKING BARRIERS, BUILDING BRIDGES

. . . .

THE CURRENT CONFIGURATION OF US Science and Technology is based on a particular model of research—"basic" research vs. "applied" research—that has significant negative effects on the practice of research. The widespread adoption of this model has created significant dissonance between what researchers actually have to do to get their jobs done, and the categories and classification systems that fund and govern their work. These classifications operate at the highest levels of government. Take for example, the congressional debate over President Obama's budget for fiscal year 2016.

As reported in February of 2015 in Science magazine,[1] Representative Brian Babin (R-TX), concerned about the requested 42 percent increase in a particular Department of Energy program—Energy Efficiency and Renewable Energy (EERE), as compared with the requested 5 percent increase for the Office of Science's "basic" research programs, asked, "Are the Office of Science's basic research programs a lower priority for this administration, when compared with these renewable programs?" Later in the same conversation, Representative Randy Hultgren (R-IL) claimed that, "There's a problem that has been ongoing with this administration's choice to value applied R&D over basic scientific research."

What is going on here? The argument is simply that by requesting a greater increase in funding for the EERE program—which provides funding for "applied" research toward new energy technologies, over the Office of Science—which provides funding for the national labs and "basic research," the Obama budget valued "applied" research over "basic" research. The Honorable Ernest Moniz, Secretary of Energy, replied by stating that the 5 percent boost for "basic" science would be sufficient and defended the so-called "applied" programs, saying "We think we have to work across the entire innovation chain."

The Republican led House of Representatives responded to the White House's request with a proposal that cut funding to the EERE program while increasing funding to the Office of Science.[2] In other words, the House of Representatives expressed a political preference for funding "basic" research over funding "applied" research. On the face of it, this is perfectly reasonable, the political process exists to allow for this sort of back and forth between the various political parties and their priorities. There is however a grave problem here, and it has nothing to do with political preferences.

Research, by its very nature, is integral. It constantly shifts between discovery and invention. Research leads to new ideas, new tools, and new devices. Classifying certain research activity as "basic" and other activity as "applied," and adopting funding models that reinforce this division is highly problematic. For one, it creates conflict by pitting "basic" research against "applied" research. This false dichotomy is taken up by the program managers who make funding decisions. These program managers require researchers to write grant applications that conform to the rationale of "basic" vs. "applied" research. Researchers in turn, organize their activities and their laboratories in such a way as to optimize their productive efforts to fulfill the stated objectives of their funded grants. As will be seen in the following discussion of certain department of energy research programs, this dichotomy can impose significant challenges to research efforts and creates unhelpful and artificial boundaries.

Upon accepting his Physics Nobel Prize in 1956, William Shockley addressed the basic/applied divide by arguing that its primary purpose is derogatory.[3] According to Shockley, terms such as "pure," "basic," and "applied" are used to elevate research driven by motivations of aesthetic satisfaction over research driven by a desire to improve a process. If the "basic" vs. "applied" model is so problematic that one of the most renowned researchers of the twentieth century would describe it as "derogatory," why is it still so prevalent and under what circumstances did it emerge to dominate policy thinking about scientific and engineering research? To begin to answer that question, we turn to the 1940s at a foundational moment for US Science & Technology.

Post-War Science and Technology Policy

At the end of World War II, the President of the United States, Franklin Delano Roosevelt, was interested in preserving the great scientific and technological efforts of the war and in turning them toward peacetime activities. He hoped the efforts that had helped stop fascism could now be harnessed to improve the national health, to create new enterprises that would lead to jobs, and to ensure the "betterment of the national standard of living."[4] President Roosevelt wrote to Vannevar Bush, Director of the Office of Scientific Research and Development (OSRD), requesting his recommendations on moving the activities of the OSRD from a wartime footing to focus on peacetime goals.

Bush's report in response to Roosevelt's letter was comprehensive. The report, *Science, the Endless Frontier* can be summed up as arguing that the best way to accomplish increased industrial research activity would be to "increase the flow of new scientific knowledge through support of basic research, and to aid in the development of scientific talent." *Science, the Endless Frontier* was completed in July 1945, a few months after Roosevelt died in office and over the next few years, as President Truman succeeded Roosevelt, the ideas contained in *Science, the Endless Frontier* became foundational to the organization of federal policy—enshrining the idea of a fundamental divide in research between "basic" and "applied"—and setting the parameters of the debate around federal funding of science and technology through the rest of the twentieth century and into the present. Bush's ideas about the relationship between "basic" and "applied" research were probably shaped by his experience of managing the OSRD during the war. Though he did not invent the terms and was not the first to divide research into "basic" and "applied" categories, the influential nature of his report ensured the widespread adoption of these silos by the federal government.

However at the same time in the 1940s, a few hundred miles north of Washington DC, at Bell Laboratories in Murray Hill, NJ, a very different model of scientific and technological research was being developed, the fruits of which would have an outsized impact on the post–war world. Imagine for a moment, that you, our reader, are faced with the challenge of rebuilding after the war. What would you have done?

Research and Solutions to Hard Problems

The year is 1947. Harry S. Truman is President of the United States. For the first time ever, the proceedings of the US Congress will be broadcast on television. 1947 is the year of the Marshall Plan—the beginning of the process of crafting a recovery for Europe after the horror of World War II. There are many problems confronting the government—from the threat of communism to the challenge of healing and rebuilding after a devastating world war. Envision that you are tasked by the president of the United States to provide funding for a significant research effort that would change the world for the better—for the sake of argument, let us assume your mission is to dramatically improve global communications in a bid to heighten understanding among the people of the world. What would you invest in? What sorts of restrictions would you place on the research that you fund?

Remember, there is no Internet, no Facebook, and no Netflix. Mobile phones do not yet exist; Twitter is something only birds do. Indeed, the entire *electronics* industry has not yet come about. There is no color television, there have only been nineteen Academy Awards ceremonies and the movie industry is not yet the juggernaut it is today. The world is a much smaller place. Your present day experience of the unique ability of modern communications technologies to collapse space and shrink distance is not yet imagined—even by the most ardent science fiction writers. Given that you, our reader, have been tasked with making this critical investment in the future by funding transformative research, who would you support? Again, what would you invest in?

Bell Labs in 1947

A very similar question faced the science and engineering leadership team at Bell Labs. Their post–war task was to continue to transform the communication industry in a fundamental way. A major challenge facing them in 1947 was to build an amplifier that would allow signals to be sent clearly over long distances. William Shockley, a director at Bell Labs, headed up a unique group that was working hard on the problem. Shockley himself had been working on a possible solution to the long distance amplifier problem for a number of years and had made some progress but was unable to build a working device. Two of his group members, John Bardeen and Walter Brattain, were the ones who made the significant breakthrough toward a solution, successfully inventing a solid-state device (an electronic device without any mechanically moving parts) that could be used to boost an electrical signal.

On December 23, 1947, Bill Shockley led Harvey Fletcher and Ralph Bown—his bosses at Bell Labs—into the solid-state research laboratory where Brattain switched on the prepared equipment and using an oscilloscope successfully demonstrated how a small bit of germanium was able to amplify an electrical signal—a feat that had, until this invention, required the use of a vacuum tube. They had invented the transistor. Showing the true utility of the device, Brattain then went on to use the transistor to amplify his voice and drive a pair of headphones—replicating the iconic moment in 1876, when Alexander Graham Bell excitedly requested Mr. Watson's presence from an adjacent room using the first-ever telephone. This first use of the transistor (right there on the laboratory bench) provided firm and conclusive evidence that the era of the vacuum tube was coming to an end.

It is clear from our collective experience in the present time that the transistor has been uniquely transformative. It has been taken up and used in multiple ways, reinterpreted and transformed in new inventions. The integrated circuit and today's iPhone contain transistors. However, it was not possible to know this in 1947. What should have guided your decision making as to the best way to spend your research dollars in 1947? In this book, we suggest that the key elements of an answer to this critical question lie in the story above.

First, Shockley's research team was fully interdisciplinary in its makeup. The team was comprised of a theoretical physicist (Shockley himself), an experimental physicist (Brattain), and a brilliant mathematical physicist and electrical engineer (Bardeen).[5] Brattain was the hands-on person, using his finely-honed intuition to experiment with various materials. Working with a superb team of Bell Labs chemists, metallurgists, engineers, and technicians, Brattain preoccupied himself with building devices. Bardeen was a whiz at theory and, working closely with Brattain on the same laboratory bench, he painstakingly collected data that he applied to explaining the phenomena they observed together. Shockley's prior attempts in the area of solid state amplification provided the team with focus and impetus.

Appreciating all this, you, our reader, would therefore be well-advised to support a cross-disciplinary research team with members capable of working well together, having complementary expertise, the very best of technical support *and* an innate desire to combine theory and practice.

Second, remember that the 1947 device (described as a point-contact transistor because it operated at a literal point) had been preceded by other failed efforts (famously, Shockley had previously worked on field-effect devices). As such, the breakthrough of December 1947 was as much a triumph of scientific understanding and theory as it was of the process of growing the necessary crystals and building the device through which the team was able to observe the transistor effect—*by actually amplifying the human voice right there at the laboratory bench*! Knowing this, you would be best advised to invest in a team capable of moving between theory and practice, between discovery and invention, between building devices and being comfortable with testing them. You would be looking for a team capable of testing its theories by rapidly building prototype devices necessary to support its work.

Third, it is clear that the Bell Labs environment itself contributed greatly to the project, not just in material resources, but also in creating the kind of intellectual environment where the transistor could be invented and discovered. As the one handing out research grants, you would therefore be interested in the institutional culture housing the research team you are handing the public's money over to. Is it supportive of collaborative, multidisciplinary teams that take an integrative approach to their work?

If the answers to all these questions are clearly positive, you are undoubtedly making a wise investment of the public's funds.

Forward to 2015

Let us jump forward in time to 2015. Here again, as in 1947, there is a major challenge—namely how do we address the problem of energy in a resource constrained world where the environmental cost of fossil fuels is so high? Unlike the thought experiment above, this is not a challenge of yesteryear. It is real, pressing, and grave. It makes sense then that a few years ago, the Department of Energy (DOE), under the leadership of Bell Labs alumnus Secretary Steven Chu, began two programs to address the multidisciplinary problem of creating solutions to future energy requirements—the Energy Frontier Research Centers (EFRC) and the Energy Innovation Hubs (EIH). The thinking was that the EFRCs would more closely align with universities and National Laboratories, whereas the EIHs would be closer to the National Laboratories and industry. The hope was that this arrangement would bridge a number of identified divides and promote interdisciplinary teams with the necessary expertise to create the needed future energy technologies and get them to market.

If we examine what is going on here in detail, it is clear that these programs were designed in ways that are resonant with our 1947 case. Both EFRCs and the EIHs were multidisciplinary, engaged a diverse range of expertise, and explicitly designed to move between theory and practice, all the while focused on future technologies. What could go wrong? What could possibly prevent them from achieving their potential?

We spoke to a number of people who were part of the review teams that gave feedback on the running of the Energy Frontier Research Centers. What we found was shocking. While it was clear that the centers were doing excellent work in materials science and engineering and making great strides in understanding the physics and the chemistry of their various projects, there was a clear and demonstrable widespread reluctance to actually build devices to prototype their ideas. Repeatedly, the EFRC directors argued that they could not build prototypes because to do so would mean that they were no longer doing "science." As the EFRCs are funded by the Office of Science in the Department of Energy, the directors concluded they could not be found to be doing work that was too "applied."

On the face of it, the irrationality of this claim is amazing, but unfortunately demonstrative of the current state of affairs. Interestingly, the researchers we interviewed admitted to using obfuscatory language in their reports and proposals in a bid to placate their program managers—for example avoiding the term "device." In essence, they were hiding their efforts to go where their research led them. Their fear of the program manager's objection to anything that wasn't considered "science" drove much of their reluctance to prototype new devices. Imagine if Brattain had been prevented from building various devices to test the ideas that he and Bardeen were experimenting with. Imagine if Brattain and the technicians at Bell Labs had been unable to make changes to their designs as he and Bardeen sat at the laboratory bench and discussed the project with Shockley. Imagine if they were prevented from demonstrating their efforts by showing, right there at the laboratory bench, that the new device was capable of amplifying the human voice? If that was the case, it is exceedingly unlikely that they would have invented the transistor or discovered the transistor effect. What if Bell Labs had encouraged its laboratory directors to monitor employees in an effort to prevent them from building prototypes at all costs (such an absurd thought!) If Bell Labs had been organized in such a way, it would never have invented the transistor and it would not have accomplished any of the other breakthroughs associated with it.

What we have in 2015 at the DOE is a research ecosystem that enshrines the distinction between "basic" and "applied" research and furthermore devotes significant resources to policing the boundary between the two. The consequence is the sad situation where researchers funded through "basic science" research dollars cannot undertake any activities that might be considered "applied science" research. This is the case even when maintaining this boundary defeats the initial mission objective of the research endeavor. In short, Secretary Chu's integrative vision was actively derailed by mid-level program managers who policed the boundaries out of concern that appropriated dollars be spent correctly. In essence, they actualized the division between "basic" and "applied" research, enforcing the distinction between the two. This absurdity describes the current state of significant portions of the publicly funded research enterprise in the United States.

The preservation of this artificial boundary between knowledge and practice is not limited to the Department of Energy. Take for example, President Obama's initiative—Brain Research through Advancing Innovative Neurotechnologies (BRAIN). This research effort was designed to "revolutionize our understanding of the human mind and uncover new ways to treat, prevent, and cure brain disorders like Alzheimer's, schizophrenia, autism, epilepsy, and traumatic brain injury" in the words of the White House.[6] The mission-oriented nature of the BRAIN research initiative is similar in many ways to the telecommunications mission Bell confronted in 1947 and to the DOE's current programs to advance new energy technologies. In this case, however, funding is being dispersed through a number of different institutions—the National Institutes of Health (NIH), the Defense Advanced Research Projects Agency (DARPA), and the National Science Foundation (NSF) with the NIH assuming the lead role.

The NIH developed a working group to act as an advisory committee to the NIH Director—the Advisory Committee to the Director (ACD). As documented in a 2013 article in *Science* magazine[7] this committee comprised fifteen neuroscientists with very little crossover between the neuroscience research community and the physical sciences and engineering research community.[8] Though the *Science* article argued for an interdisciplinary perspective and even raised the question of whether or not recent technological developments render the BRAIN imitative itself unnecessary, it is clear from the final composition of the ACD that the NIH's BRAIN program remains firmly oriented in neuroscience. The ACD membership did not involve any active researchers in associated primary disciplines such as computer science, electrical engineering, or physics. Where is the integration?

The ACD Brain's working group report, *BRAIN 2025: A Scientific Vision*, was published in June 2014 and contains many laudable big tent arguments. However, it also clearly defines the NIH's BRAIN program as a neuroscience effort, and though it goes as far as discussing the need for interdisciplinary efforts toward a "concerted attack on brain activity at the level of circuits and systems,"[9] our reading of the entire episode is that it represents a missed opportunity to fully engage with this challenge in a transformative way, as the effort remains firmly rooted in preexisting frameworks of the neuroscience community and creates stove pipes between different federal agencies. The two-mission-oriented research funding agencies (DOE and NIH) have segregated their research efforts. Indeed, undertaking truly integrative transdisciplinary efforts has been a challenge for the NIH and DOE as documented recently.[10]

Toward a More Holistic View of Research

The challenges outlined above can be described as a clash of models. On the one hand, we have an integrative and holistic model of research—best exemplified by Bell Labs. It draws no distinction among research activities but rather recognizes the need for a porous boundary between research and application. On the other hand, we have a model, outlined in the pages of *Science, the Endless Frontier*, which imposes rigid boundaries between particular research activities described as "basic" and "applied" and devotes significant resources to maintaining this boundary—even when the natural inclination of researchers is toward research activities that resist this simplistic classification.

We call the integrative model the discovery-invention cycle, and have based it upon the research environment at Bell Laboratories and some of the other great industrial research laboratories which flourished in the United States over an extended period of the twentieth century. In this book, we lay out the specifics of the discovery-invention cycle and contrast it with the "basic" vs. "applied" model. We discuss how the "basic" vs. "applied" model creates and reinforces artificial boundaries in science and engineering, in research and translation, and in discovery and invention. What we are interested in is investigating various institutional arrangements and practices that promote the engagement of knowledge and practice, catalyze trans- formative correspondence between different domains of knowledge and expertise, and act in such a manner as to have a real and posi- tive effect on science and engineering research environments in the United States. What can be done to build different kinds of research institutions along the lines of the discovery-invention cycle?

As a consequence of this broad question, this book operates at intersections. The intersections of science, engineering, and technol- ogy; of theory and practice; academia; the national laboratories; and industry; science and technology studies (STS); and public policy. We explore some of these intersections and associated, relevant, research institutions. We draw upon Narayanamurti's widespread experience of operating at these intersections and at a number of critical institu- tions from academia to private research institutions, public research institutions, public-private hybrids, and at various levels of policy making. Odumosu's hybrid training as an engineer and STS scholar has allowed us to integrate what is known from the social sciences that could be adapted in various ways to address institutional design and policy making. It is this rich tapestry of experiences that informs the arguments in these pages.

A few common terms used frequently throughout the book de- serve some exposition here. They include: Research, which we un- derstand as an *unscheduled* quest for new knowledge and the creation of new inventions, whose outcome cannot be predicted in advance, and in which both science *and* engineering are essential ingredients. Development, which is a *scheduled* activity with a well-defined out- come in a specified time frame, aimed at the marketplace. Discovery, is the creation of new knowledge and facts about the world, and Invention, is the accumulation and creation of knowledge that re- sults in a new tool, device, or process that accomplishes a particular or specific purpose. We understand Innovation to encompass discov- ery, invention, research and development, extending these into the products, processes, and ideas that result in significant improvements in the world.

In the next chapter we discuss some persistent boundaries in re- search and in science and technology. We show how widespread boundary drawing is and provide some historical context for the formation of these problematic distinctions. Chapter 3 undertakes a review of the evolution of the basic/applied framework and argues that it is no longer useful. Even though US science and technology policy has relied on this framework for much of the last six decades (for organizing research activity, creating and maintaining funding institutions, and valuing particular types of research over others), it is clear that new and fresh thinking is needed to adequately respond to future challenges. We present our rationale for the need to jettison the basic/applied framework and show that it is not reflective of actual research practice.

In the fourth chapter we examine the history of the basic/applied framework and show how this way of thinking about science and technology research became widespread in the halls of Congress and in official US Science and Technology policy. The fifth chapter presents our alternative to the basic/applied framework. We expand upon our model—the discovery-invention cycle and argue for a more holistic view of science and engineering research based on a detailed examination of how research actually proceeds. In the sixth chapter, we discuss how this view of discovery and invention would play out in an institutional setting. What would it meant to run a research institution along the lines of the discovery-invention cycle? The research environment at Bell Laboratories is to our minds, the clearest example of a discovery-invention research environment in practice. The sixth chapter is devoted to examining the culture of Bell Labs with a mind toward identifying critical elements of institutional research culture.

Having identified the critical institutional elements necessary to cultivate the discovery-invention cycle, in Chapter 7, we investigate the possibility of building new kinds of research institutions along similar lines and discuss the commitments that are required to build them. The seventh chapter is written as a case study of two exemplary contemporary institutions that have successfully integrated discovery-invention thinking into their design. Chapter 8 steps back from examining individual institutions and asks what changes are necessary in national research policy to promote the widespread adoption of discovery-invention style institutions and move away from the basic/

applied dichotomy. Chapter 9 presents our concluding arguments and summary.

We will not attempt a comprehensive analysis of the entirety of the US Science and Technology policy ecosystem. Such a review would require several books. This book is also not meant to be an exhaustive look at the entirety of the research environment. Rather, we limit ourselves to disciplines with which the authors are familiar and then extend our core argument to other spheres of knowledge. At its core, this book is about transforming US Science and Technology policy away from an obsession with policing the boundary between "basic" and "applied" research toward a more integrative vision of research as a holistic process. We acknowledge that the boundary-work[11] that goes into sustaining the status quo is intellectual as well as political[12] and any intervention that seeks to move the conversation along has to engage with both the intellectual and political dimensions of the current shape of US Science and Technology policy. We believe this book lays out a clear intellectual argument for change and provides a blueprint to effect it.

2

BOUNDARIES IN SCIENCE AND
ENGINEERING RESEARCH

. . . .

FOR MANY, ENGINEERING IS understood as "applied science" yet, upon closer inspection, we find that many engineers are engaged in activities that are self-evidently science. The most prestigious award in science, the Nobel Prize, has been awarded to several engineers. In 2000, Jack S. Kilby, an electrical engineer, won the Nobel Prize in physics for his part in the invention of the integrated circuit. Charles Kao, an electronics engineer, won the 2009 Nobel Prize in physics for his work on optical fiber communication. Similarly, scientists have also won Nobel engineering prizes. T. Peter Brody, a physicist, won the National Academy of Engineering's Charles Stark Draper Prize in 2012 for his work on liquid crystal displays. In 2006, Willard Boyle and George Smith, two physicists, won the 2005 Draper Prize for their work on charge-coupled devices. Yet, engineering and science have different social capital associated with them. In many arenas "scientist" is a more prestigious moniker than "engineer." But what does the distinction between science and engineering really describe (beyond a convenient means of group identification), if the work of practitioners of both can be recognized at the highest levels as either science or engineering?

A similar sort of distinction exists in descriptions of research as "basic" or "applied." Take for example the Nobel Prizes, long acknowledged as the epitome of scientific (read "basic research") achievement. It is interesting that the Nobel Prizes were instituted by a man who, although trained as a chemist, is perhaps most accurately described as an engineer and an inventor.[1] Furthermore, the Nobel Prize is itself sustained by a fortune that was made as much from perspicacious patents and shrewd business practices, as it was from the invention of dynamite.[2] In many ways, Alfred Nobel was prototypical of the great inventors—Thomas Edison, Alexander Graham Bell, and George Westinghouse—and his historical trajectory mirrored theirs much more than the lives of Einstein, Bohr, or Newton.

Considered as a whole, however, these great pioneers of science/engineering were individuals whose own work, experience, and history demonstrated to them the importance of integrated research practices. Nobel's fortune, for example, was built on a mixture of inventiveness and business acumen that enabled him to leave it to honor those who "during the preceding year, had conferred the greatest benefit on mankind."[3] Edison's inventiveness, business, and managerial genius resulted in not only over a thousand patents and multiple inventions, but also in the creation of Edison Electric's Menlo Park, the first industrial research laboratory in the United States.[4] Menlo Park's culture of integrative research and prodigious output would later be duplicated in the General Electric Laboratories. Alexander Graham Bell's seminal inventions in telephony which led to the creation of the American Telephone and Telegraph Company (AT&T) and the creation of its research and development arm—Bell Laboratories— would give the world the transistor, the charge-coupled device (CCD), and information theory. Furthermore the organizational structure of the old Bell Laboratories is still celebrated as a prime example of the successful melding of "basic" and "applied" research, integrating the work of scientists and engineers. Indeed much of the impetus behind this book comes from Venkatesh Narayanamurti's, (one of the authors of this volume), experiences at Bell Labs directing integrated research and hiring and facilitating exceptional researcher practitioners, some of whom later went on to win Nobel Prizes.

Notwithstanding these historical examples of holistic approaches to research, it is clear that the basic/applied binary continues to be widely institutionalized in the design of US national research enterprises.[5] Consider for example, the well-known rivalry between the national security research laboratories. While Los Alamos and Lawrence

Livermore National Laboratories were meant to provide competition for each other in the areas of physics of nuclear weapons development, the role of Sandia National Laboratories has been that of developing the nonnuclear components of the weapons systems, in other words, the "engineering" or "applied" lab to the "science" or "basic" labs at Livermore and Los Alamos. Not only are the functional responsibilities of the three labs organized around this basic principle of dividing the "science" from the "engineering," the culture and environment of each of the labs reflects this split. The governance of Sandia for much of its history was entrusted to AT&T, while the management of Livermore and Los Alamos fell to the University of California.

Knowledge Boundaries and Educational Institutions

The persistence of these boundaries can also be observed in our elite educational institutions. As C. P. Snow pointed out in his well-known 1959 lecture on the two cultures, the elite Universities do not have a record of supporting the teaching of the more "technical arts." Oxford, Harvard, and Yale, for example, have historically had very little engineering education in their curricula. Harvard's creation of the School of Engineering and Applied Sciences is a recent intervention that is itself an echo of a much earlier Harvard institution, the Lawrence Scientific School and the bequeathal of the great American inventor, engineer, and entrepreneur—Gordon McKay. McKay's bequest was specifically given for supporting scholarship in the applied sciences (engineering).[6] The tale of engineering at Harvard is a complicated one that begins in the early 1900s with then Harvard President Eliot's attempt to merge Harvard and MIT. The Lawrence Scientific School was about to receive the McKay bequest and President Eliot convinced President Pritchett of MIT to merge MIT with the Lawrence Scientific School over the objections of both Harvard and MIT faculty. The proposed merger was blocked on a technicality used by the courts to prevent MIT from selling its Boston property.[7] Again and again, Harvard courted the idea of establishing an engineering school, and as the McKay bequest was specific about how the funds were to be used, the new school would have to be an integral part of Harvard's liberal education. The rise of information technology and the advent of the Internet, coupled with the creation of Microsoft by Harvard alums Bill Gates and Steve Ballmer, led to an increased acceptance of the importance of disciplines such as computer science and electrical engineering in the early 1990s.[8]

However, it wasn't until the mid-1990s that a concerted effort, led by then Harvard President Rudenstine and Dean Knowles of the Faculty of Arts and Sciences, succeeded in attracting Venkatesh Narayanamurti to move to Harvard and to accept the challenge of building an engineering school at Harvard, where he subsequently developed the notion of the Harvard renaissance engineer. At the launch of the new school, President Charles Vest of the National Academy of Engineering and Emeritus President of MIT pointed out the importance of engineering at Harvard, as symbolizing to the country the new relevance of technology becoming a part of liberal education.

Boundaries, Categories, and Definitions

Boundaries are not impermeable. A lot of work is required to police and sustain their borders. This intellectual as well as political boundary-work is an effort to sustain the status quo. Any intervention that seeks to move the conversation along has to engage with both the intellectual and political dimensions of the current shape of US Science and Technology policy.

One key insight is that the language and definitions that structure US Science and Technology policy evolve and change over time. This has been well documented by historians of science and technology. While ideas and definitions of "basic," "applied," "pure," and even "science" and "engineering" are not static, the resources and commitments at stake in these definitions and assumptions behind them are significant. A consequence is that the S&T policy status quo has become so divorced from actual practice that in many cases it is now an impediment to the research process.[9] Vannevar Bush described all the engineers working in the Office of Scientific Research and Development (OSRD) as "scientists" in order to "counter what he saw as the American military's antipathy towards engineering salesmanship and British snobbery toward engineering" (Kline 1995, 219). The interplay between the profession and the professional and the blurred lines between them are clearly demonstrated in the discussion of the origins of Maxwell's equations below.

Maxwell's Equations, Physics, and Electrical Engineering

Research in the physical sciences has long been of an integrated nature with theory and application working hand-in-hand. The contemporary divide separating physics and electrical engineering is relatively new. Many of the subjects discussed in physics and electrical engineering were once the purview of the natural philosophers. While institutional and professional identities have contributed to the maintenance of the boundaries between physics and electrical engineering, it is important to remember that disciplines are fluid (as evidenced by birth and death of new and old disciplines and the existence of hybrid disciplines like biochemistry) and their boundaries are porous. Significant portions of electrical engineering and physics are highly related, and the research trajectories in either field can intersect and bisect each other. One clear example of this kind of integrated research can be found in the work of James Clerk Maxwell.

Though Maxwell's well-known equations are sometimes held up as an example of how theoretical science can provide tangible benefits, as the historian Simon Schaffer has shown,[10] the eponymous equations were actually catalyzed by the need to solve a classic engineering problem, namely the necessity of improving signal propagation in undersea communication transmission lines.

Interested in the problem of signal delay in undersea telegraph lines, Maxwell on the advice of his mentor Lord Kelvin spent years working with observed data from undersea cables, using the engineering data to refine his model of electromagnetics. When the first attempt at a transatlantic undersea cable ended in disaster, in a bid to tackle the problem and as a result of the high economic value of a workable solution, British scientists including Maxwell gained access to even more precise electromagnetic measurements. It was the availability of these values that led to the realization that electrostatic and electromagnetics units were equivalent. This fundamental insight allowed Maxwell to integrate the formerly separate areas of light and electromagnetism with profound effects for the future of physics *and* of electrical engineering.

In the preface to his 1873 work *A Treatise on Electricity and Magnetism*, Maxwell himself acknowledged the importance of engineering, and the usefulness and utility of accurate electrical engineering measurements (which were generated in attempts to solve engineering challenges) to the development of his thinking and life's work. In fact, the current modern form of Maxwell's equations was actually published in 1885 in *The Electrician*, a journal for electrical engineers, in an article authored by Oliver Heaviside.[11] Heaviside, a self-taught electrical engineer and mathematician, was interested in minimizing losses in electrical transmission and came up with the current modern notation of Maxwell's four equations. Heaviside's work and his interpretation of Maxwell's equations was instrumental to the implementation of the telegraph.[12] Even in this early example the cyclical nature of discovery and invention is clear in the interplay between the discovery of new knowledge, represented by Maxwell's equations, and the engineering challenges in signal transmission related to the invention of the telegraph.

It is apparent then that the natural sciences and engineering have long been interlinked and that practical pursuits have never been far from the acquisition of new knowledge. However, in contemporary US science and technology policy and in the governance of many of our mission-oriented research agencies, this simple idea has been forgotten and silos and boundaries have built-up, impeding the progress of knowledge and inventions . . . the status quo is unsustainable and unacceptable. The solutions to the pressing problems facing the nation are too reliant on science and engineering for us not to ensure that they proceed interactively. This is the important message that we hope you will take away from this book. In the next chapter we turn to a thorough examination of one of the key boundaries in US science and technology policy—the boundary between "basic" research and "applied" research.

3

THE BASIC/APPLIED DICHOTOMY

The Inadequacy of the Linear Model

. . . .

THE CURRENT BOUNDARY BETWEEN "basic" research and "applied" research has had and continues to have a significant deleterious impact on US science and technology policy. As was discussed in the first chapter, DOE middle level program managers defeated the integrative efforts of stated DOE policy because they were constantly policing this boundary. They were obsessed with ensuring that work funded with "basic" dollars not be labeled as "applied" and vice-versa. This mind-set leads to absurdities of the type that have already been described. What drives these individuals, who are well educated and rational, to extremes of this sort? The problem lies in our understanding of what "basic" research is, what "applied" research is, and how this understanding propagates from the halls of the US Congress to the laboratory bench of funded engineers and scientists. In essence, the DOE program managers discussed in the first chapter were effectively being guided by an accounting mentality which requires rigid separation into silos, rather than Secretary of Energy Chu's vision and the spirit of the Energy Frontier Research Centers (EFRC) and the Energy Innovation Hubs (EIH).

The important question is, what causal model of research and industrial activity underlies this division of research into the silos of "basic" and "applied?" What model of innovation contains the notion of "basic" vs. "applied" research as a core element? Perhaps more important, can a model of innovation that pits "basic" vs. "applied" research truly succeed?

A good place to look for answers to these questions is in the history of the linear model of innovation. This model holds that innovation is a linear process that begins with basic research, proceeds through applied research, product demonstration and deployment, and then delivers a product, good, or service that is taken up by business and introduced into the marketplace. The basic/applied framework can thus be considered as representative of the first two stages of the linear model of innovation.

Basic Research → Applied Research → Product Development → Market Diffusion and Deployment

In this book we are more concerned with the aspects of the linear model that deal with research. While the steps required to introduce technologies into the marketplace successfully are important, of greater interest to us are: the implied relationship between science and technology in the language of "basic" and "applied" research; the processes of discovery and game changing inventions; and the order in which they occur.

The linear model in its simplest, distilled essence, represents the belief that the technological products of modernity are driven by a process that begins with basic (sometimes referred to as fundamental) research, which is then taken up in further research activities that aim to solve specific problems, the outcomes of which may lead to the development of products and services that may prove to be marketable. The line that leads from the laboratory bench to the marketplace lends the linear model its narrative. It tells a story of how science and technology work, and provides clear levers for policy intervention. The logic of the linear model suggests that to increase the quantity of innovative products all that need be done is increase the previous activities on the line (i.e., fund more basic research). The notion of linear progression from the lab bench to the market suffuses science and technology policy at the highest levels of US government and policy making. One need look no further than President Obama's 2012 State of the Union where he claimed that innovation "demands" basic research. The State of the Union address is useful generally, reflecting large guiding assumptions that inform both behavior and policy. In that speech, the president drew a causal link between innovation, which impacts society broadly, and basic research, then tied both to national competitiveness. The lab benches in universities are directly and necessarily linked to, for example, new cancer treatments and lightweight vests for cops and soldiers.

There is no doubt that this accounting of the working of science, engineering, and technology is at times correct, but it is at best a very partial and incomplete picture of how the science and technology enterprise functions. It is clear, though, that there is widespread adoption of the linear model in federal science and technology policy and in the design of federally-funded research institutions. Much of this widespread acceptance of linear model thinking in US public policy has a single, influential source. It can be traced to Vannevar Bush's 1945 report to President Roosevelt—*Science, the Endless Frontier.*[1]

Vannevar Bush and *Science, the Endless Frontier*

In his letter to Vannevar Bush that resulted in the report *Science, the Endless Frontier,* Franklin Delano Roosevelt, the President of the United States, requested that Bush focus on four specific goals. First, the president was interested in making the knowledge generated during the war available to the world. Roosevelt clearly spelled out his belief that the "diffusion of such knowledge should help us stimulate new enterprises. . . ." drawing a clear connection between the spread of new knowledge and economic activity. Second, Roosevelt wanted to know how the scientific resources of OSRD could be turned toward the "war of science against disease." Third, the president wanted clear advice on the best way for the government to promote research by both public and private organizations. Finally, and related to the third point, President Roosevelt asked Vannevar Bush to develop a program for discovering and nurturing scientific talent in the United States on a similar level as achieved during the war.

Bush's report focused on science and scientific research arguing that the promotion of research activity would be in the public good and meet the needs set out by President Roosevelt. It is interesting to note that although he was trained as an electrical engineer and was a former professor and dean of engineering at MIT, in his iconic report Bush hardly uses the term engineering.[2] Instead, the report is a strong statement advocating the importance of science and the need to promote and protect it. It argues for a cordoning off of "basic" from "applied" research. In a striking section titled "The Importance of Basic Research," the report defines "basic research" as research "without thought of practical ends" which leads to general knowledge about nature and its laws. It contrasts that with "applied research," which is described as focused on solving practical problems. The report argues that "basic research" *may* not always lead to practical problem solving, but still it generates new knowledge—"scientific capital" from which fund practical applications *must* be drawn. It states that "basic research" is the "pacemaker" of technological progress and argues that the funding of such research is in the clear national interest. It is in sections like this, that the rationale of the linear model is clearly spelled out. Also, by omitting the role and importance of engineering Bush conflated science with "basic research," leaving out the critical element of "engineering research." To make his point, Bush cites a number of examples of successful wartime projects and attributes their success to science. One of the most important of these contributions extensively referenced in his report was the rapid development of radar technology.

Radar allowed the allies to "see" objects at sufficient distances and granted them a significant edge in their ability to respond to air, water, and submarine attacks. The development of Allied radar capabilities was a major contributor to the final successful outcome of the war.

Engineering and Winning World War II

That one of Vannevar Bush's principle examples of scientific enterprise during the war is radar is ironic, for it dramatically makes clear that it wasn't just "scientific research" of the "basic variety" that was critical to the war effort. After all, using radio waves for detection and ranging had been going on for a while and the principle of radar was well established. Rather, it was the invention of a device to increase the power of radar systems to a practical, usable level that ensured the success of allied radar development.

The invention of the cavity magnetron in the United Kingdom, which was later given to the United States in what has been described as the most important reverse lend-lease item of all time,[3] was crucial to the success of radar technology. In essence, it was the *engineering* of an important *technological invention*—the cavity magnetron—that allowed for the successful use of radar.

The term radar was initially an acronym for Radio Detection and Ranging (RADAR) that has since become a common noun—radar. Radar systems are predicated on the simple observation that given knowledge of the speed traveled between two points and the time of departure and arrival, distance can be easily calculated. In radar systems, electromagnetic waves are reflected off objects, and the arrival time of the detected reflection is measured, allowing for a calculation of distance (range). Sufficiently complex radar systems can also be used to create maps and topographies of faraway objects.

The basic principle behind radar was known long before World War II. The science behind it was well understood—especially in the UK where it was used in the early warning air defense system known as Chain Home. However, the UK's radar, and other radar systems used by other allied forces suffered from a number of limitations. The early systems utilized waves of longer length and as a result, their directionality and resolution (ability to discriminate detail) was relatively poor. The size, complexity, and energy requirements of these early systems also made mounting them on mobile platforms a challenge. It was clear that in order to improve the resolution, directionality, and mobility of radar systems, it was important to build systems that used much shorter wavelengths (microwaves). The challenge was producing these waves continuously at high power in a device that would not be too unwieldy to transport.

At the University of Birmingham, in Professor Mark Oliphant's physics laboratory, John Randall (a physicist who had previously worked for General Electric) and Henry Boot (a postdoc at Birmingham) developed just such a device, the multicavity magnetron. Their invention integrated the best elements of the klystron (a coherent high frequency amplifier device developed by Russell and Sigurd Varian at Stanford

University) and the magnetron (a radio frequency oscillator) allowing them to build a device that generated high power microwaves. Their invention influenced the war effort in no small way.

What is clear is that Randall and Boot's contribution is best described as a newly-engineered invention that integrated aspects of prior inventions. While Randal and Boot were both physicists, their work in creating a new tool for the generation of high-power microwaves was surely more an engineering intervention aimed at solving a practical problem than "pure" science. Yet, nowhere in Bush's account of their accomplishment, was this critical point made clear.

Of course, the importance of engineering to the successful prosecution of the war effort was not limited to Radar alone. Engineering played a critical role in the Manhattan Project, in the development of synthetic rubber, the proximity fuse, and other ordinance innovations that led to the Allies' technological superiority and their victory over the Axis Powers. The war was won not only through the efforts of science but rather, through a combination of exciting engineering innovations and work in the physical sciences.

A close and detailed examination of the arguments made in *Science, the Endless Frontier*, where Bush credits Basic research with being essential to critical successes in the war, while omitting direct mention of engineering's contributions, leads to a much more complicated picture. In almost all instances, there was as much engineering as there was science in the work of the OSRD. Why then was engineering given such scant attention in *Science, the Endless Frontier?* In some regard, the rationale of the linear model that places emphasis on science over engineering builds on some long-standing ideas in Western culture (and other cultures as well) that favor the work of the "head"—or intellect—over that of the "hand"—including in this case technical expertise.[4] A central contention in this book is that it is more productive to examine the creative power of the imagination to transform both "hand" and "head" activities.

The Linear Model in Other Contexts

While Vannevar Bush's *Science, the Endless Frontier* was largely responsible for the wide-scale adoption of the linear model in US science and technology policy, the roots of the linear model predate the report. The historian of economics, Benoit Godin, argues that the linear model developed through three stages.[5] In the first stage (roughly 1900–1945), the US National Research Council and business leaders, including those involved in corporate research labs, began focusing on the relationship between "pure" and "applied" science, arguing that the former preceded the latter. In the second phase, the notion of development, which arose out of the business literature of the 1920s, became a part of the linear model. Proponents of development argued that taking steps to get an object or idea to market was just as important as the original research. Finally, in the last stage, thinkers in business schools and other spaces concerned with innovation began applying the linear model to a wide variety of activities, "like production and diffusion."[6] Godin argues that the "linear model owes little to (Vannevar) Bush" but that "it is rather a theoretical construction of industrialists, consultants and business schools, seconded by economists." This is a point with which we somewhat agree. It is clear that Bush did not invent the dichotomies he so masterfully employed. Indeed, some of the ideas about the divisions between basic and applied research are actually in Roosevelt's letter to Bush![7] However, it is also clear that Bush's treatment of the relationship between "basic" and "applied" research and his argument for the primacy of basic research played a crucial role in institutionalizing linear model ideas and concepts in the federal government and in the policy-making process.

Reexamining the Direction of Causality

The linear model postulates a particular causal direction. In the world of the linear model, basic research always precedes applied research, implying that science comes before engineering. However, the question of whether understanding always precedes invention has long been a troubling one as, for example, in the 1960s when historians of technology began to differentiate themselves from historians of science. The history of science, which had its primary roots in the beginning of the twentieth century, tended to describe technology as "applied science." In the 1970s, the historian of technology,

Edwin T. Layton, Jr., published a pair of essays that fundamentally questioned the linear model's validity. In his 1971 essay, "Mirror-Image Twins: The Communities of Science and Technology in 19th Century America," Layton argued that the scientific and technical (especially engineering) professions had grown up along parallel paths during the late nineteenth century.[8] During this time, engineers were not subservient to scientists; they did not "apply" the knowledge of scientists; rather, they had their own professional organizations and institutions apart from the world of science. In another essay, "Technology as Knowledge," Layton went further, asserting that technology (or more accurately, technological design) was its own form of knowledge that could intersect with science but did not necessarily do so.[9] Many forms of technology reached relatively advanced stages of development before detailed scientific explanations about how the technologies worked emerged. In one of the most famous examples, James Watt invented his steam engine before the laws of thermodynamics were postulated. As is widely quoted, "the science of thermodynamics owes more to the steam engine, than the steam engine owes to science."[10] In later chapters, we will examine several examples of this kind of reversing of the directionality of the linear model.

Moving Beyond the Linear Model

The question remains, where do we go from here? How do we produce a positive vision of how to shape science and technology (instead of merely a negative project of criticizing an old view)? Some steps toward a new vision have been taken.

During the 1970s and 1980s, alternative models, for example, the National Innovation Systems (NIS), demarcated a move away from pipeline models of innovation like the linear model. The economist Richard Nelson described a National Innovation System as "a set of institutions whose interactions determine the innovative performance of national firms."[11] The National Innovation System model operates on the assumption that innovation, and the technologies that emerge from it are the result of complex interactions among key institutions. These institutions, and the ways and manner in which they are arranged, are key to promoting innovation. Relevant institutions include firms, research universities, industrial laboratories, and government laboratories. NIS marked a turn toward institutions and was part of a broader trend—the "new institutionalism" that was popular at the time within the humanities and social sciences, including economics and political science. The trend signaled a desire to move away from abstract social theory to one that embraced the messiness and murkiness of history and organizational change. The NIS literature created a nuanced vision of innovation that still retains power today, and promoted a more holistic view of the intertwining of science and technology, however, it did little to undermine the picture of the science-technology relationship inherent in the linear model.

It wasn't until 1997 when Donald Stokes' *Pasteur's Quadrant: Basic Science and Technological Innovation* was published posthumously that any serious modification of the linear model began to permeate US science and technology policy. In his book, Stokes argued that scientific efforts were best carried out in what he termed "Pasteur's Quadrant." If the linear model is imagined as a line, then Stokes' quadrants are achieved by expanding to two dimensions and including an axis that considers the applicability of research efforts. Pasteur's Quadrant is then the quadrant that enjoys both high levels of fundamental knowledge and high levels of applicability—as exemplified in Pasteur's work on vaccination and consequent contributions to microbiology. Innovation in Pasteur's Quadrant is that which is motivated simultaneously by expanding understanding and also improving our abilities (technological, including medicine) to change the world.

Stokes' experience as a dean at the University of Michigan and at Princeton University inspired his book. He often experienced scientists talking about research in ways that were, as Stokes put it, "odd and helpful."[12] The publication of Stokes' book was itself hailed as finally laying the vagaries of the linear model to rest once and for all. A blurb on the back of the book quotes US Congressman George E. Brown, Jr., stating that "Stokes' analysis will, one hopes, finally lay to rest the unhelpful separation between 'basic' and 'applied' research that has misinformed science policy for decades." But, while clearing the ground for future research, Stokes did not go far enough.

Stokes viewed the linear model as a statement about motivations or intentions. Stokes work indicates that he saw the relevant questions as: What is driving researchers in particular cases? Is it a desire to contribute to scientific knowledge (basic research)? Is it a wish to improve technology (applied research)? Or is it perhaps some kind of hybrid of the two (Pasteur's Quadrant)? This focus on the researcher's motivations did nothing to address the preserved boundary that remains implicit in his model. Indeed, his very nomenclature reinforced the binaries of the linear model. Pasteur's quadrant became the quadrant of "use-inspired basic research"—still compatible with the linear model binary of "basic" and "applied" research.

Another way of interpreting the linear model, however, is to view it as a statement about which factors go into producing innovation. By factor, we simply intend its common meaning: "a circumstance, fact, or influence that contributes to a result or outcome," as the dictionary states, although we should also note that the absence of something can be as powerful a factor as the presence of others. From this perspective, the linear model asserts that technical change requires whatever factors make up "basic science," and as long as we ensure that these factors are present and healthy—through government funding and such—we will have the result or outcome, namely technological progress.

Stokes makes some steps toward questioning this picture. For instance, in his representation of the "revised dynamic model," he notes that both "existing understanding" (e.g., science) and "existing technology" contribute to the hybrid space of "use-inspired basic research."[13] This view has been useful in somewhat destabilizing the entrenched position of the linear model and has led recently to a small shift in the understanding of the relationship between basic and applied research. This shift is evident, for example in the report by the National Academies, "Rising above the Gathering Storm" and the American Academy of Arts and Sciences report, "Advancing Research in Science and Engineering II (ARISE II): Unleashing America's Research & Innovation Enterprise."[14]

Yet, while Stokes notes how "often technology is the inspiration of science rather than the other way around," his revised dynamic model does not recognize the full complexity of innovation, preferring to keep "basic" and "applied" in their own paths that only mix in the shared agora of "use-inspired basic research."[15] It is also instructive to note that Stokes' framework *preserves the language of the linear model* in his use of the terms basic and applied as descriptors of research.

Why was Stokes somewhat successful? Stokes was not alone in critiquing the linear model. While sociologists, historians and other social scientists likewise voiced criticisms, their critiques have not been taken up by the policy world or by scientists and engineers.

This is an important question and can only really be answered by a thorough empirical study, however, some portion of Stokes' success is due to the usefulness of simple models for making sense of the research process. The linear model for all its reductionism, lends itself to simple analysis and the identification of possible levers for policy intervention. Stokes' revised dynamic model also presents a simple enough quadrant-based tool for policy communication. If our assumption is correct, Stokes modest dent in the linear model's continued relevance serves to explain, in some small part, the inability of policy discourse to engage with the social science work on refuting the linear model. After all, if we do away with the linear model, what do we replace it with that can act as a kind of schematic shorthand for discussing the research process? What tools and frameworks can we draw on to find alternative conceptions of the relationship between "basic and applied research" and engineering, and the nature of radical shifts in science and technology?

A New Vision for Science and Technology

Some other initial steps toward a new vision have been taken. For example, Hendrik Casimir, a distinguished Dutch theoretical physicist and leader of the Phillips Eindhoven Laboratory, proposed the idea of the science-technology spiral.[16] In his model, Casimir describes science and technology as two separate "streams" where technology always draws upon scientific results and science depends on technology. In Casimir's model, technology never uses the most recent or profound results of science, science always uses the most recent and advanced technologies. Technology utilizes science, albeit with a time lag, while science always uses the latest technologies with no time lag. His science and technology spiral can fairly be characterized as firmly in the mind-set where technology is always and merely an "applied science." While a number of physical sciences researchers[17] have found some utility in the idea of the spiral as a "virtuous circle", Casimir's model has found little audience among social scientists and policy makers, leaving Stokes' model as the dominant variation of the linear model to enjoy broader uptake.

However, Stokes attempts to complicate the linear model preserves the language of basic and applied research. While Stokes' quadrant system expands the binary of basic and applied, it does little to address the implied hierarchy between science and engineering, and does not deal at all with the reality of newly created knowledge being an essential element of every technological advancement.

Perhaps more important, the effectiveness of research cannot be fully appreciated by a singular attention to the motivations of the researchers involved. Motivations, of course are important. Working toward finding a cure for cancer or to advance the frontiers of communications can be a powerful incentive to stimulate ground breaking research. That said, we must also take into consideration other important dimensions of research. Broadening our scope of inquiry involves not only examining researchers' motivations but also examining how the research project is accepted, implemented, and disseminated, and how it brings together the essences of science (discovery) and engineering (creation). An important implication of this long-term view is the need to encourage the inventive spirit that leads to the creation of new technologies.

After a clear examination of all of these aspects of the research process—researchers initial and concluding motivations, an appreciation of how the research project is accepted, implemented, and disseminated, consideration of the long-term view, examining how the research catalyzes other research and the need to nurture future technologies—the limitations of the basic/applied binary enshrined in the linear model become painfully obvious.

To overcome these limitations, we propose using the terms "invention" and "discovery" to describe the twin channels of research practice. For us, the essence of invention is the "accumulation and creation of knowledge that results in a new tool, device or process that accomplishes a particular, specific purpose." The essence of discovery is the "creation of new knowledge and facts about the world." Considering the phases of invention and discovery along with research motivations and institutional settings enables a much more holistic view of the research process. This allows us to examine the ways in which research generates innovation and leads to further research in a truly virtuous cycle.

Research in practice is a complex, nonlinear process. Still, and particularly for policy-making purposes; straightforward and sufficiently realized representations like Stokes' Pasteur's Quadrant are useful analytical aids. With both facts in mind, we propose the discovery-invention cycle which will serve to illustrate the interconnectedness of the processes of invention and discovery, and the need for consideration of research effectiveness over longer time frames. This model will be fully explained and developed in the following chapters. The model will allow for a more reliable consideration of research innovation through time. The model should aid in discerning possible bottlenecks in the functioning of cycles of innovation, indicating possible avenues for policy intervention.

In summary, we are advocating a shift from thinking about research in terms of motivations (basic vs. applied) alone to a broader consideration of the (to borrow a term from Layton[18]) "mirror-image twins" of discovery and invention. We also believe that research should be evaluated not only by its initial motivations, but also by its ability to catalyze other research and innovations. Our studies suggest that such a rethinking is necessary for the creation of useful policy and the design of more effective research institutions—where long time frames, a premium on futuristic ideas, and feedback between different elements of the research ecosystem are essential ingredients.

The discovery-invention cycle is a model that avoids the problematic characteristics of the "basic" / "applied" framework. We will expand upon this model in Chapter 5. Before we undertake this, it is important to understand the historical context that gave rise to the widespread adoption of the "basic" / "applied" framework. The next chapter examines this origin.

4

THE ORIGINS OF THE "BASIC"
AND "APPLIED" DESCRIPTORS

GIVEN THAT DIVIDING UP research along the lines of "basic" and "applied" is injurious to science and technology, an important question needs to be addressed. Where and how did this distinction emerge? What conditions led to the current "basic" vs. "applied" framework, and how do we escape it?

As we argued in the previous chapter, Vannevar Bush's description of "basic" and "applied" research did much to shape the evolution of science and technology in the post–war United States. The influence of *Science, the Endless Frontier,* especially in the widespread take up and adoption of the term "basic research" used in the way Bush described, is clearly seen in the Congressional Record where usage of the term spiked concurrently with the publication of the report. Bush's report resulted in a number of influential changes in the US government's approach to research funding, and though his proposed National Research Foundation (NRF) that would centralize federal funding of basic research never materialized, in many ways the National Science Foundation (NSF) proposed by Senator Harley Kilgore of West Virginia reflected Bush's ideals of basic research funding. Unlike the unrealized NRF, the NSF was only one of many federal institutions that funded both "basic" and "applied" research. The messy politics of the post–war period resulted in today's ecosystem of multiple and overlapping research funding institutions, including the Office of Naval Research, the

National Institutes of Health, and the Department of Energy, to name but a few others. The multiple funding paths and redundancy helps guard against groupthink and allows for a more robust research environment as the various agencies adopt different research design perspectives. Nevertheless, it is clear that the various federal agencies internalized the core message of *Science, the Endless Frontier*—namely the need to separate "basic" research funding from "applied" research funding and application. Bush's definition of "basic research" as research "performed without thought of practical ends" effectively enshrined the hard line between basic and applied research funding throughout this ecosystem.

The Changing Nomenclature of Research 1880–2000

The terms "basic research" and "applied research" have been the subject of much of this book's critique so far and our intention is to hasten their demise. A parallel effort is a much needed reconsideration of how we imagine the process of research along the lines of the discovery-invention cycle which we will do in the forthcoming chapters. Taken together, these efforts would result in a more effective research policy environment that is better aligned with the natural course of research practice. However, before we move to considering alternative models of research policy it is useful to examine how the terms "basic" and "applied" emerged to their current level of prominence and the antecedents of the current descriptors.

As Ronald Kline, the historian of science and technology has shown, for centuries there has been a raging debate over the hierarchy and status of various fields of learning. There was a time in the nineteenth century when subjects that are currently considered the mainstay of theoretical physics (optics and acoustics, for example) were once dismissed as "mixed mathematics" as opposed to "pure mathematics."[1] An examination of the historical record shows clearly that the distinction between fields of knowledge accorded respect and fields of knowledge that were looked down upon is actually quite permeable. Of course, change doesn't happen on its own. A number of fascinating historical characters were instrumental in changing popular views of different fields of knowledge-making.

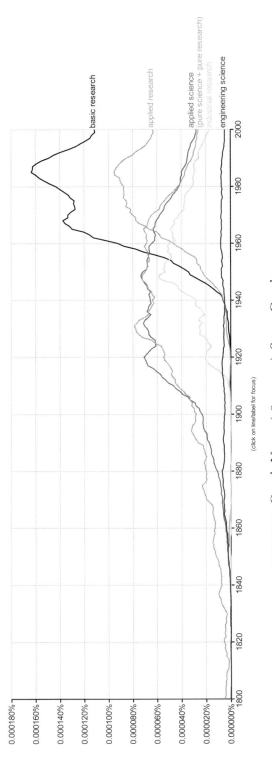

FIGURE 4.1. Google Ngram (1800–2000). *Source:* Google.

As is clear in Figure 4.1, the modern terms "basic research" and "applied research" only really began to achieve widespread use in the late 1940s—this nicely correlates with the publication of *Science, the Endless Frontier*. However "basic" and "applied" gained in popularity at the expense of other descriptive terms like, *pure science, fundamental science, applied science,* and *industrial research*. It is useful to review the trajectory of these descriptive terms and to examine some of the commitments they represent.

Pure Science

In the 1870s in the United States, the *pure* in *pure science* was a comment on the motives of researchers. Anyone undertaking pure science was presumed to have pure motives for undertaking research. At heart, purity meant a research pursuit for knowledge's sake alone with little to no consideration for application or, God forbid (almost literally), wealth. It went without saying that doing anything else was "not" pure. The ideal of *pure science* is perhaps best represented by the physicist Henry Rowland. His famous 1883 speech, "A Plea for Pure Science" laid out the ideals behind the *pure science* label.[2] That speech was really a reaction to the prominent status of the great inventors like Edison whose work on electricity was being called "physics" in the popular press. Rowland did much to promote science over engineering and technology, though his undergraduate degree was ironically in engineering.[3]

Applied Science

Applied Science, unlike pure science (which has by and large retained its meaning), has at various times meant different things to different constituencies. As Kline points out, one of the early uses of the term was by engineers in reaction to Henry Rowland's *pure science* ideal. A president of the American Society of Mechanical Engineers, Robert Thurston, began to lecture widely about *applied science*. Thurston's speeches reflect a wide range of interpretations as to the meaning of *applied science*. His usage of the term ran the gamut from *applied science* being a particular and different kind of *engineering* knowledge to *applied science* being the application of science to the "useful arts." Along with the electrical engineer, Charles Steinmetz, Thurston promoted engineering as *applied science* with its own research practices and customs in line with the scientific method, as well as an activity that utilized scientific knowledge for practical purposes. This sophisticated view of engineering has at times been replaced with a simple utilitarian view of *applied science* as being merely the application of science.

Industrial Research

The term *industrial research* has largely remained descriptive of the kinds of research activity that take place in industrial laboratories. However, beyond its descriptive interpretation, it has played a large role in problematizing the meanings and interpretations of *pure* and *applied science*. Especially during the war, industrial research labs were places where proponents of both *pure* and *applied science* mixed. Industrial research during the war required the expertise of both scientists and engineers. Their working closely together on common projects led to some broadening of opinions. The urgency of war has proven to help clarify the benefits of less doctrinaire divisions. After the First World War, for example, Robert Millikan concluded that the "distinctions between the man whom you commonly call the pure scientist and the man whom you commonly call the applied scientist have absolutely disappeared."[4]

Basic and Applied Research

As shown above, the modern framework of *basic* and *applied research* really began to take off after the publication of *Science, the Endless Frontier*. Interestingly, before the publication of the report, passages that described some research as *pure* research were amended and the term *basic* was utilized in place of the descriptor *pure*. Bush's decision to make substitutions in the various committee reports that made up the final report came about partly in response to rhetorical shifts that took place after the war, but also because of criticism by Frank Jewett, the first president of Bell Labs (see Kline 1995). Jewett was concerned that the use of "pure" implied that other kinds of research were "impure" (see Kevles 1995, p. 45). In spite of Jewett's expressed desire to avoid value judgments that would privilege certain types of research activity over others, the "basic research" vs. "applied research" dichotomy was established and the language of the report led directly to the linear model that still bedevils our science and technology policy communities. Another consequence of ascribing most of the successes of the war effort to science was the continued dominance of physicists in the setting of science and technology policy after World War II.

The publication of Vannevar Bush's report also provided simple criteria to judge research and determine the boundaries of particular research activities, that is, "Is the research of the basic variety?" If not, then it must be of the other kind. This kind of simplistic evaluative standard is a boon to bureaucratic systematization and rationality, and resulted in just the kinds of stove piping any student of the social theorist Max Weber[5] would expect. In the paper that directly led to this book[6] we argued that when examined from the perspective of researchers, this distinction between basic and applied is not helpful and can even be disruptive to the practice of research. As discussed in Chapter 1, imagine if someone had told John Bardeen, Walter Brattain, and William Shockley that they could not build the bipolar contact transistor because doing so wouldn't be "basic" research? Never mind that it was the very invention that contributed to their discovery of the transistor effect; a clear example of the union of theory and practice! This isn't as much of a stretch as one might imagine. In our current era, conversations regularly take place between researchers and funding managers as to how "far" they can take their work and still have it qualify as "basic" research.

The Struggle between Research Categories and Research Reality

In recent correspondence with the authors (see below), Dr. Lyle Schwartz, formerly the Director of the Air Force Office of Scientific Research, and Director of the Materials Science and Engineering Laboratory at the National Institute of Standards and Technology, neatly captures how high the stakes have grown due to policy makers' embrace of "basic" vs. "applied" science. Dr. Schwartz discussed[7] his lengthy career in government service and identifies some of the challenges the basic/applied distinction imposed on the functioning of the National Institute of Standards and Technology (NIST) and the Department of Defense (DoD) categorization of different phases of research and development. For the DoD, Basic Research (6.1) leads to Applied Research (6.2) that results in Advanced Technology Development (6.3)—the numbering system encodes the linearity of the model very clearly, indeed.

Consider these circumstances I lived:

The DOD S&T system does use bins for R&D. The 6.1, 6.2, and 6.3 categories rather rigidly constrain how much money can go to work of different degrees of "advancement." Furthermore, this work is characterized by readiness "levels" first instituted by NASA, Technology Readiness Levels (TRL), and more recently, for development and beyond, Manufacturing Readiness Levels (MRL). I directed AFOSR, funded almost exclusively by 6.1 money and therefore, supposed to do "basic" research (TRL 1–2). Of course, AFOSR was also supposed to be an integral part of the Air Force Research Lab which meant that ~30 percent of our funding was to be spent within the lab by AFRL scientists and engineers.

This meant that in planning our portfolio of activity, in executing at the lab, and at universities in the United States and abroad, and in evaluating the impact of our work, we needed to describe our work in terms that would enable our senior leaders to appreciate why they were spending these dollars. This meant linking our funded work to possible implications if successful and properly transitioned to the next level of activity.

As can be seen from the excerpt above, the work of NIST, while requiring some aspects of "basic" research, was simultaneously judged on the basis of "improved measurement and standardization." This dual evaluative system resulted in struggles by NIST researchers to explain and justify their important work—this was the case even though the institute's work has been recognized with a number of

Nobel and Draper Prizes in addition to many distinguished commendations. Schwartz and his colleagues were delighted with the expanded language of Pasteur's Quadrant as it helped explain the integrated nature of their work. However, as can be seen by the lack of change in DoD nomenclature, even the limited intermingling of notions of basic and applied represented by Donald Stokes' Pasteur's Quadrant has barely worked its way into the calcified institutional categories and definitions of the US Federal government.

The challenges of explaining and representing the work undertaken at AFOSR and NIST that Schwartz describes are a clear byproduct of the rigid boundaries that the basic/applied framework requires. Equally clear is that the actual practice of research is a multifaceted and contingent process and resists the simple ideal categories of "basic" and "applied."

Schwartz's embrace of Stokes' solution is understandable. However, Stokes' solution to the limitation of the dual categories of basic and applied research was an increase in the number of categories from two—basic/applied—to three—pure basic research, use inspired basic research, and pure applied research (or four if you count the unnamed quadrant).We will not rehash our critique of Stokes and the limitations of his model presented in Chapter 3. What we are interested in here is an exploration of the various ways in which the basic/applied framework has affected US S&T policy and various attempts to resolve the tensions between the framework and the actual practice of research. To do this, we will examine three important tenets in *Science, the Endless Frontier* that have become an integral part of the social contract between the research community, the state, and its publics. First though, it is useful to examine the social and political policy environment in which Bush (an electrical engineer) was writing and examine the cultural resources he relied on to make his arguments. To do this effectively requires a quick review of modern American industrial history.

The Policy Dominance of the Great American Industrialists

The late 1800s were an amazing period for American industry. This period marked the emergence of the great inventors and industrialists and the world-changing industries that they pioneered: namely, Thomas Edison, Alexander Graham Bell, and Henry Ford. All household names, each invented and transformed important industries— electrical power, communications, and transportation—in addition to pioneering new industrial processes and forms of production. During this time, American scientific research was seen as a distant second to European science research. European scientists published more papers, won most of the Nobel prizes, and Europe was still the place to go for graduate studies in science. Engineering, however, was everywhere gaining in popularity with rising university enrollments, driven by the increasingly complex needs of industry. Yet, science originating in America was rapidly growing up and improving in stature and reputation, particularly with Millikan's success at the California Institute of Technology (Caltech). The country's primary scientific association, the American Association for the Advancement of Science (AAAS), created in 1848, was by design elitist and modeled on analogous European institutions such as the Royal Society. It did not admit engineers, inventors, or industrialists. Nevertheless, most of the transformative inventions that were perceptibly changing the world were traceable to the work of engineers, inventors, and industrialists. Even more troubling for some of these industrialists was the fact that in the popular media their work was being described as "science" and the result of "physics."

As we described earlier, in a response to what he saw as the cheapening of science, Henry Rowland, an unabashed advocate of "pure science" and a defender of science elitism against the problematic equalizing notions of "democracy," gave his famous speech—"A Plea for Pure Science." Rowland, professor of physics at Johns Hopkins and at the time in 1883, vice president of physics at the AAAS, argued that physics needed to be enshrined as a science rather than "call telegraphs, electric lights, and such conveniences by the name of science."

Despite Rowland's call for pure science, senior policy makers remained enamored of the great inventors and industrialists and their views were held in high regard. This period, before the onset of World War I can be thought of as the age of the great inventors and the veneration of the practical arts. Science, and research more broadly, was about to take its place on the national stage, and it all began with the start of hostilities in 1914.

The war in Europe began having a perceptible and negative effect on commerce and industry in the United States with the British Blockade that prevented the delivery of key industrial supplies such as dyes, fertilizers, and scientific instruments. The lack of these critical components made the dearth of American capabilities clear to all and led Willis R. Whitney, the director of the General Electric Research Laboratory, to argue in the pages of *Science* magazine (which began life as a magazine financially supported by Thomas Edison and Alexander Graham Bell), that research needed to become a national duty, and should be seen as a "necessity to any people who are ever to become a leading nation or world power."[8]

As the war began to loom ever larger, Thomas Edison took to the pages of the *New York Times* to assure the nation that American cleverness and ingenuity would prevail over the German U-boats.[9] The interview was read avidly by Josephus Daniels, the secretary of the navy, and in response he instituted the Naval Consulting Board (NCB) overseen by Edison himself, and the entirety of its membership derived from industry and the inventor class. Edison advised the secretary that he did not think that any scientific research would be necessary to the work of the NCB. The board focused instead on tapping engineering talent to address problems, and it was widely seen as a national recognition of the importance of engineering as a profession.

The NCB had only two "men of science" and they were mathematicians from the American Mathematical Society. One of these gentlemen, Arthur Gordon Webster unsuccessfully attempted to convince the secretary of the navy to appoint National Academy members and a few members from the American Physical Society. At the first meeting Webster was informed that Edison desired to have the Board composed of "practical men who are accustomed to doing things, and not talking about it."[10] Such was the cultural status of the great inventors and their sway over national policy. The clear dominance of engineering and the industrialists caused Gregory Hale, distinguished astronomer, foreign secretary of the National Academy of Science, and editor of the *Astrophysical Journal*, to write of the need to raise the status of physics to the point where it might "penetrate to the sanctum of the secretary of the Navy."[11]

Hale, who was politically astute, worked hard to convince President Wilson of the importance of research for defense and his efforts lead to the creation of the National Research Council (NRC) to encourage both "pure" and "applied" research for national security. Hale gained the support of both political parties and firmly established the NRC's role in government. When the United States severed diplomatic ties with Germany on February 3, 1917, the NRC was tasked with the responsibility of finding a way to detect German U-boats.

Hale and the physicists he enrolled to assist him (Millikan, in particular) were immensely successful in this and at other efforts that the NRC undertook. Meanwhile, though Edison and the NCB proposed about forty-five devices for the military, the navy ignored them all. The result of this experience was that to the military, it became clear that (as Kevles has argued) the advance of defense technology required the organized efforts of scientists and engineers whose first steps often had to be, as Admiral Griffin said of submarine detection, "in a sense, backwards in the unexplored regions where fundamental physics truths and engineering data were concealed."[12]

The success of the NRC effectively marked the end of one era and the ascendancy of scientific research, particularly physical science research in defensive and offensive weaponry. This ascendancy was cemented by World War II, which has been described as the physicist's war, especially with the success of the Manhattan Project. Both wars ensured the unchallenged cultural status of science in general and physics in particular.

Consequently, the science and technology culture in which Vannevar Bush found himself was one dominated by a near veneration of science, especially in the aftermath of the atomic bomb. Engineering had had its heyday and science was in the enviable position of supreme cultural authority. This goes a long way to explaining why Vannevar Bush so clearly aligned himself with "science" when he wrote his report, even though as exemplified in the radar case discussed in Chapter 3, much of the research activity that won the war was clearly carried out by engineers—including Bush himself. Armed with this historical context, we can proceed with a close reading of some of the tenets found in *Science, the Endless Frontier.*

Three Tenets in *Science, the Endless Frontier*

The first deserving attention is the notion that technology and innovative products are dependent on "basic" research.

> "new products and new processes do not appear full-grown. They are founded on new principles and new conceptions, which in turn are painstakingly developed by research in the purest realms of science." (Bush, p. 19)

Here, in very clear language, Bush argues that new products and processes (technology) are dependent on pure scientific research. This is a simple restatement of the linear model that we have confronted previously, but it is important to highlight the clear logic of the linear model in Bush's report and the ways that "basic research" in "the purest realms of science" is held up as being in the public interest. According to Bush's formulation, it is this that leads to new principles and new conceptions. This often repeated argument is invoked along with references to the distinguished economists Joseph Schumpeter and Robert Solow. However, a close reading of both Schumpeter and Solow reveal that neither supported the linear model[13] nor argued for its saliency.[14]

The second tenet we will consider circumscribes the appropriate role for government, that is, the funding of basic research, and read in parallel with the prior tenet above, it also identifies the *limits* of government intervention.

> a nation which depends upon others for its new basic scientific knowledge will be slow in its industrial progress and weak in its competitive position in world trade. . . . Industry is generally inhibited by preconceived goals, by its own clearly defined standards, and by the constant pressure of commercial necessity. Satisfactory progress in basic science seldom occurs under conditions prevailing in the normal industrial laboratory. There are

some notable exceptions, it is true, but even in such cases it is rarely possible to match the universities in respect to the freedom which is so important to scientific discovery. (Bush, p. 19)

Accordingly, in Bush's estimation, government funding should focus on basic research of the sort that industry is unable to undertake, and the fruits of this labor will result in greater industrial activity. Bush thus dismissed the research conducted in industry as primarily "applied research," though today we would probably recognize much of what went on in industry during much of his time as "basic research." It was as groundbreaking (perhaps even more so) as work done in universities and other research environments.

For example, Clinton Davisson's 1937 Nobel-winning work on electron diffraction undertaken at Bell Laboratories would definitely qualify in Bush's framework as "basic research" as would Irving Langmuir's 1932 Nobel-winning work on surface chemistry carried out while at General Electric. Finally, consider Karl Jansky's seminal engineering research in 1933 at Bell Labs that did as much as anything else in birthing the field of radio astronomy. Jansky, known as the father of radio astronomy, is a textbook case for jettisoning the basic applied dichotomy. Trained as a physicist, Jansky was hired by Bell Laboratories to investigate noise in radio transatlantic transmission. He built an apparatus to solve this most practical of problems and discovered different sources of noise, one of which he correctly identified as a signal emanating from the center of the Milky Way. This discovery, catalyzed by an investigation into the practical problem of noise in communication circuits, birthed a whole new field of science—radio astronomy—that utilizes radio waves to study the universe.

All of these took place in industrial research laboratories and even using Bush's framework definitely qualify as "basic research activity." These examples of ground-breaking research make it even more baffling that Bush was so dismissive of the work done at Industrial Laboratories. As we will see later in this book, there was a proliferation of such examples in several industrial laboratories after World War II.

The third tenet of Bush's report establishes the need to protect "pure research" from "applied research."

"a perverse law governing research" under which "applied research invariably drives out the pure." (Bush, p. 83)

Bush's "perverse law governing research" provides the rational basis for much of the institutional boundary drawing (stove piping) in federal funding and in the organization of research activity. For, if applied research poses a threat to "basic" research, it makes sense that a function of the bureaucracy should be to ensure that "applied" research is not *allowed* to encroach upon "basic" research. This perceived need to isolate the two has led to a number of absurd policing efforts by concerned bureaucrats. Perhaps more problematically, it has also led to self-disciplining of researchers, where funded researchers limit the natural outcomes of their research activity and rhetorically contort their reporting to ensure that the boundaries between "basic" and "applied" research are maintained.[15]

Drawing the Line Appropriately

In contrast, our argument continues to be that research activities are more accurately described as a cycle of discovery-invention rather than by a fixation on the basic/applied dichotomy. Such an approach is, we believe, appropriate for the public good.

Ask any researcher who works in a laboratory and you will quickly discover that initial motivations are a poor predictor of the ultimate end-point of their research activity. Motivations are fuzzy, and are hardly ever only about knowledge, or only about possible applications. Discovery, and the creation of new knowledge often catalyzes new applications. In the same vein, the creation of a new invention, a new engineered material, or a new device, can often lead to the discovery of new facts about the universe. The basic/applied dichotomy in its insistence that research be segregated along the lines of knowledge and application is a flawed model that does not reflect how science and technology advance through laboratory research.

A salient feature of research is the need to protect it from short-term pressures. Research by its very nature is a long-term effort, whose outcomes cannot be predicted with certainty. This is equally true for the quest for new knowledge and for identifying engineered solutions to pressing societal challenges. Bush attempted to accomplish this by dividing research into two varieties, "basic" and "applied." We contend that this was a mistake due to the cyclical nature of the processes of invention and discovery. The line should not have been drawn between varieties of research, since all forms of research are about the unknown. The line should rather have been drawn between *research*, which needs to be an unscheduled activity irrespective of its domain—science, engineering, or technology—and *development*, which is about the process of creating a specific product in a specified time frame and is therefore subject to the imperatives of the marketplace.

Another important point is about scale of the resources devoted to these activities. In the Bell Labs case, for example, resources expended on development were at least an order of magnitude greater than what was spent on research, and the boundary between the two was porous by design. The most useful value here is *insulation* not isolation. There needs to be leakage and cross-talk between the two. Individual researchers need to have the flexibility to move across the boundaries between research domains, and in certain cases, between research and development—when it is productive to do so. In the next chapter we provide clear examples of these principles in practice in our detailed exposition of the discovery-invention cycle.

Having concluded our discussion of the shortcomings of the "basic"/"applied" framework we can now undertake a thorough exposition of an alternative model of research activity and organization namely, the discovery-invention cycle. The next chapters provides a full treatment of the discovery-invention cycle along with numerous illustrative examples of how it could be adopted in policy making and the design of research organizations.

5

THE DISCOVERY-INVENTION CYCLE

· ·

THE DISCOVERY-INVENTION CYCLE (DIC) is a model of research[1] that addresses many of the problems identified in Pasteur's Quadrant, namely its retention of the classification of research into two categories of basic and applied, it's singular focus on researcher motivations, and its inability to capture the contextual and relational nature of research. It is our contention that the narrow identification of discovery research as the wellspring of technological progress paints an incomplete picture. The DIC requires the recognition of the interdependency of discovery and invention. The challenge lies in expanding the discourse beyond a simple acceptance of discovery research incorporating the equally important but often neglected role of inventions in generating both new knowledge (discoveries) and new applications. Most people intuitively understand these concepts but the bidirectional and interdependent nature of the two isn't usually acknowledged.

To illustrate the power and utility of the DIC consider Figure 5.1, where we trace the evolution of the technologies underlying the current information and communication age. What can be said about the research that has enabled the recent explosion of information and communication technologies? How does our model, the DIC, enable a deeper understanding of the multiplicity of research directions that have shaped the current information era? To fully answer these questions, it is necessary to examine research over time, paying attention to the development of knowledge and the twin processes of invention and discovery. Most important is tracing their interconnections through time. The awarding of the Nobel Prize offers excellent snapshots illustrating the interaction of invention and discovery that enabled the information age.[2]

Further illustrating the interconnectedness of science and engineering, some of the events described next were also commemorated by the Draper Prize of the US National Academy of Engineering. The Draper Prize has been described as the "Engineering Nobel" lending an elegant symmetry to the analysis. We describe these kinds of clearly intersecting awards as a "family" of prizes in that they are all closely related. We have identified other such families whose innovation cycles can be clearly described and illustrated through time.[3]

The birth of the current information age can be traced to the invention of the transistor. This work was recognized with the 1956 physics Nobel Prize awarded jointly to William Shockley, John Bardeen, and Walter Brattain "for their researches on semiconductors and their discovery of the transistor effect." Building upon early work on the effect of electric fields on metal semiconductor junctions, the interdisciplinary Bell Labs team built a working bipolar-contact transistor and demonstrated the transistor effect through their discovery. This work and successive refinements enabled a class of devices that successfully replaced electromechanical switches, allowing for successive generations of smaller, more efficient and more intricate circuits. While the Nobel was awarded for the discovery of the transistor effect, the team of Shockley, Bardeen, and Brattain had to invent the bipolar-contact transistor to demonstrate it. Their work was thus of a dual nature, encompassing both discovery and invention.[4] The integrated nature of their advance can be seen in its consequences. The invention of the transistor catalyzed a whole body of further research into semiconductor physics and the invention of the bipolar-contact transistor led to a new class of devices that effectively replaced vacuum tubes and catalyzed further research into, and invention of new kinds of semiconductor devices. The 1956 Nobel is therefore exemplary of a particular kind of knowledge making that affects both discoveries and inventions. We call this kind of research, radically innovative. In Figure 5.1, the 1956 prize is situated at the intersection of invention and discovery and it is from this prize that we begin to trace the innovation cycle for the prize family that de-scribes critical moments in the information age.

The second prize in this family is the 1964 Nobel Prize, which was awarded to Charles Townes, Nicolay Basov, and Aleksandr Prokhorov. The prize recognized work that is fundamental to the modern com-munications age. Most of the global communications traffic is carried by transcontinental fiber optic networks, which utilize light as the signal carrier. Townes' work on the simulated emission of microwave radiation earned him his half of the Nobel. This experimental work showed that it was possible to build amplifier oscillators with low noise characteristics capable of spontaneous emission of microwaves with almost perfect amplification. In the course of Towne's experi-ments, the maser effect—"microwave amplification by the stimulated emission of radiation" was observed. Later, Basov and Prokhorov, along with Townes, extended that discovery to consideration of its application in the visible spectrum and thus the laser was invented. Laser light allows for the transmission of very high power pulses of light, at very high frequencies, and is crucial for modern high speed communication systems. In short, the 1964 Nobel acknowledges crit-ical work that was simultaneously discovery (the maser effect) and invention (inventing the maser and the laser), both central to the rise of the information age. In Figure 5.1, the 1964 Nobel is also situated at the intersection of Invention and Discovery. The work on lasers built directly upon previous work by Einstein, but practical and oper-ational masers and lasers were enabled by advancements in electronic switches made possible by the solid state electronics revolution that began with the invention of the transistor.

Although scientists and engineers conducted a great deal of foun-dational work on the science of information technology in the 1960s, the next wave of Nobel recognition for this research did not come until the 1980s. Advancements in the semiconductor industry led to the development of new kinds of devices such as the Metal Oxide Semiconductor Field Effect Transistor (MOSFET). The two dimen-sional nature of the MOSFET's conducting layer provided a conve-nient avenue to study electrical conduction in reduced dimensions. While studying the Hall Effect (an effect that describes the electric voltage produced across a current carrying wire when it is subjected to a perpendicular magnetic field) in two-dimensional systems, von Klitzing discovered that his empirical measurements were exact in a way that was surprising. Von Klitzing had discovered that the Hall Effect is quantized in two-dimensional systems, such as those found in highly refined transistors, in the presence of strong magnetic fields, and at very low temperatures. Instead of regular, continuous changes, von Klitzing observed steps and plateaus. The discovery was extremely useful in establishing electrical standards, as the values observed were so precise and accurate they could be expressed as integer multiples of two fundamental constants and be used to define resistance. The quantized Hall Effect can thus be seen as an important discovery with useful applications and is situated in the discovery half of Figure 5.1. It was also very important in catalyzing research into two-dimensional systems. For his work on the quantized Hall Effect, von Klitzing was awarded the 1985 Nobel.

The 2000 Nobel Prize was awarded jointly to Zhores Alferov and Herbert Kroemer, and Jack Kilby. Kroemer and Alferov got half of the Nobel for "developing semiconductor heterostructures" and Jack Kilby got the other half for "his part in the invention of the integrated circuit." The 2000 Nobels can be classified primarily as inventions. The work on heterostructures built on the previous work of Shockley et al.; again, we label these Nobels a "family" because of their interdependence. This research enabled a new class of semiconductor devices that could be used in high-speed circuits and opto-electronics. Early transistors were not fast enough for use in high frequency circuits. Alferov and Kroemer showed that creating a double junction with a thin layer of semiconductors would allow for much higher concentrations of holes and electrons, enabling faster switching speeds, and allowing for laser operation at practical temperatures. Their invention produced tangible improvements in lasers and light emitting diodes. It was the work on heterostructures that enabled the modern room temperature lasers that are utilized in fiber-optic communication systems. We should note here that although Alferov and Kroemer's work on heterostructures was recognized with the 2000 Nobel prize, the actual work was carried out, as discussed later, in the 1950s and 1960s and the consequences of this invention eventually led to the discovery of a new form of matter.

In 1989, Jack Kilby's work on integrated circuits at Texas Instruments earned him half of the inaugural Draper Prize. Kilby and Robert Noyce, cofounder of Intel Corporation and Fairchild Semiconductor Corporation, received the Draper Prize "for their independent developments of monolithic integrated circuits." In 2000, Kilby's work, again done at Texas Instruments showing that entire circuits could be realized with semiconductor substrates earned him his half of the Nobel. Shockley, Bardeen, and Brattain had invented semiconductor-based transistors, but these were discrete components, used in circuits with components made of other materials. The genius of Kilby's work was in realizing that semiconductors could be arranged in such a way that the entire circuit, not just the transistor, could be realized on a chip. This invention of a process of building entire circuits out of semiconductors allowed for rapid economies of scale, bringing down the cost of circuits, and further research into process technologies allowed escalating progress on the shrinking of these circuits so that in a few short years chips containing billions of transistors were possible. The two inventions commemorated with the physics Nobel Prize of 2000 can be traced directly to the work carried out by Shockley and colleagues. In Figure 5.1, the research honored by the Nobel Prize of 2000 is thus situated firmly in the Invention category.

Influenced by Alferov and Kroemer's prior work, and with advancements in crystal growth techniques, (Molecular Beam Epitaxy which allowed for atomically precise layered heterostructures), Stormer and collaborators invented the concept of modulation doping, where the charge carriers were physically separated from their parent donor atoms. This allowed for the fabrication of two dimensional electron layers with mobility orders of magnitude greater than in Silicon MOSFETs. Stormer and Tsui then studied the unusual two-dimensional electrical conduction properties of these structures. They cooled heterojunction transistors to a fraction of a degree above absolute zero and exposed them to very strong magnetic fields. By doing so they discovered a new kind of particle that appeared to have only one-third the charge of the previously thought indivisible electron. Laughlin showed through calculations that this observation was of a new form of quantum liquid where interactions between billions of electrons led to swirls that behaved like particles with a fractional electron charge. Clearly a new discovery, this beautiful phenomenon influenced by previous inventions, held important practical applications (for example, high frequency transistors used in cell phones). For their work, Robert Laughlin, Horst Stormer, and Daniel Tsui were awarded the 1998 Nobel Prize in physics. In Figure 5.1, it is situated firmly in the Discovery category, but its origins clearly proceed from a prior invention.

The 2009 Nobel Prize recognized inventions that were previously honored by the Draper Prize. Charles Kao, Robert Maurer, and John MacChesney received the 1999 Draper Prize "for development of fiber-optic technology." One-half of the Nobel Prize of 2009 was awarded to Charles Kao for "groundbreaking achievements concerning the transmission of light in fibers for optical communication" and one-half was awarded jointly to Willard Boyle and George Smith "for the invention of the imaging semiconductor circuit—the CCD." Both prizes were directly influenced by previous inventions and discoveries. Kao was primarily concerned with building a workable waveguide for light for use in communications systems. His inquiries led to astonishing improvements in glass production as he predicted that glass fibers of a certain purity would allow for long distance laser light communication. Of course, the engineering of heterostructures that allowed for room temperature lasers was critical to inventing the technologies of fiber communication. Kao, however, not only created new processes for measuring the purity of glass, but also actively encouraged various manufacturers to improve their processes. Kao's work, built on the work by Alferov and Kromer, enabled the physical infrastructure of the information age. Kao's fellow recipients of the Draper Prize applied his work in industrial settings. Maurer developed optical fibers at Corning. At Bell Labs, MacChesney enabled the mass-production of commercially viable optical fibers. Willard Boyle and George Smith received the 2006 Draper Prize for their work on Charge-Coupled Devices which was also recognized by the other half of the 2009 Nobel Prize in physics. Boyle and Smith continued the tradition of Bell Labs inquiry. Adding a brilliant twist to the work of Shockley and colleagues on the transistor, they designed and invented the charged coupled device (CCD), a semiconductor circuit that enabled digital imagery, and later on video.

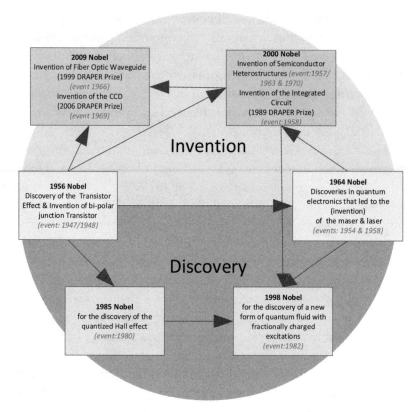

FIGURE 5.1. The Innovation cycle in information and communication technologies. *Source:* Authors.

These six Nobel Prizes and the associated Draper Prizes highlight the multiple kinds of knowledge that play into the innovations that have enabled the current information and communication age. From the discovery of the transistor effect that relied on the invention of the bipolar junction transistor, and led to all the marvelous processors and chips in everything from computers to cars, to the invention of the integrated circuit that made the power of modern computers possible while shrinking their cost. The invention of fiber-optics combined with the materials engineering and invention of heterostructures made the physical infrastructure and speed of global communications networks possible. In fact, the desire to improve the electrical conductivity of heterostructures led to the unexpected

Table 5.1. ICT prizewinners and their employers while conducting the prize-winning research.

1956	John Bardeen: Bell Labs	Walter Brattain: Bell Labs	William Shockley: Bell Labs
1964	Charles Townes: Columbia University (Bell Labs alum)	Nicolay Basov: P.N. Lebedev Physical Institute	Aleksandr Prokhorov: P.N. Lebedev Physical Institute
1985	Klaus von Klitzing: University Würzburg and High Magnetic Field Laboratory, Grenoble		
1998	Robert Laughlin: Bell Labs, Lawrence Livermore Laboratory	Horst Störmer: Bell Labs	Daniel Tsui: Bell Labs
2000	Zhores Alferov: Physico-Technical Institute	Herbert Kroemer: RCA Labs, Varian Associates	Jack Kilby: Texas Instruments *Recipient of the inaugural 1989 Draper Prize with Robert Noyce of Fairchild Semiconductor, whose 1990 death precluded the Nobel*
2009	Charles Kao: Standard Telecommunications Laboratories *Recipient of the 1999 Draper Prize with Robert Maurer of Corning and John MacChesney of Bell Labs*	Willard Boyle: Bell Labs *Recipient of the 2006 Draper Prize with George Smith of Bell Labs*	George Smith: Bell Labs *Recipient of the 2006 Draper Prize with Willard Boyle of Bell Labs*

discovery of fractional quantization in two-dimensional systems and a new form of quantum fluid. Each of these could probably be classified as "basic" or "applied" research, but such a classification elides the complexity and multiple nature of the research previously described. Worse, it reinforces the prejudices of many against what is now labeled as "applied research." Thinking in terms of invention and discovery through time helps reconstruct the many pathways that research travels in the creation of radical innovations.

In our model, the discovery-invention cycle can be traversed in both directions, and research knowledge is seen as an integrated whole that evolves over time as it traverses the cycle. The bidirectionality of the cycle captures the fact that inventions are not always the product of discovery, but can also be the product of other inventions. Simultaneously, important discoveries can arise from new inventions. The time dimension is captured in the idea of traveling the cycle in various directions.

Noble Laureates and the Industrial Laboratories

Industrial labs were the backbone of the information and communications technologies family of Nobel Prizes. Of the sixteen laureates, only six conducted their prize-winning work outside of an industrial setting. Three of these six individuals worked in Soviet research institutes. Of the remainder, one was at a German university and a German research institute, one was at an American research university, and one was at an American government laboratory.

All the American laureates had industrial research experience, particularly at Bell Labs. Affiliates and alumni were represented in more than half of the prizes overall and 70 percent of the prizes were awarded to laureates employed by industrial laboratories. Charles Townes and Robert Laughlin, the two American laureates who did not conduct their prize-winning research in industrial research laboratories, were both alumni of Bell Labs. Although Townes had left the employ of Bell Labs when he became a professor at Columbia, he continued to consult at Bell Labs.[5] Laughlin actually began his Nobel Prize-winning work at Bell, surrounded by semiconductor experts who introduced him to the potential of these materials. However, even for the Bell Labs theorists of that time, Laughlin was too unconventional and as a result his post-doctoral position was not converted into a permanent appointment, necessitating his move to Lawrence Livermore Laboratory.

Researchers outside of the industrial labs were very cognizant of the science and engineering occurring within them. At the High

Magnetic Field Laboratory in Grenoble, Klaus von Klitzing collab-
orated with Siemens Forschungslaboratorien, which helped provide
MOS-devices. Over a decade earlier, Zhores Alferov had told his
Soviet colleagues that other researchers recognized the implica-
tions of heterostructures for semiconductor physics and electrons.
By 1968, he felt that he was in a race with the research arms of three
large American companies as his main competitors: Bell Telephone,
IBM, and RCA.

Alferov's fellow recipients of the 2000 prize highlight the broad
spectrum of companies interested in groundbreaking research
and development. Herbert Kroemer worked at both RCA and
Siemens. Kroemer began his award-winning work at the Central
Telecommunications Laboratory (FTZ) of the German postal service, and moved to RCA Laboratories However, it was not until he
started working at Varian Associates that he was successful. His paper
was initially rejected by *Applied Physics Letters* and only later published
in *Proceedings of the IEEE*, during which time Kroemer filed for and
received a patent. However, the support from Varian Associates ended there. Unlike their earlier involvement and foresight with Felix
Bloch,[6] they failed to anticipate the applications of Kroemer's work
and refused him further resources.

Jack Kilby spent most of his career at Texas Instruments. Trained as
an electrical engineer, he nonetheless believed it was important to understand physics and took several engineering physics courses. These
proved useful when the invention of the transistor rendered much of
his engineering training with vacuum tubes obsolete. Kilby began his
career at the Centralab Division of Globe Union, Inc. working on
transistors under a license from Bell Labs, where he foresaw the scale
of resources necessary to compete in the race for miniaturization.
While IBM, Motorola, and Texas Instruments all made him offers,
their proposals differed. The team at IBM was working in a direction
that Kilby thought would prove unfruitful. Motorola was interested
in miniaturizations but proposed that he only work part-time. Texas
Instruments, however, seemed to be the most enthusiastic and offered
to let him work on his project full-time as a special assignment, which
paid dividends. Prototypes of what would become the integrated circuit rapidly followed. Part of this was due to Kilby's deduction that

higher overheads at Texas Instruments meant that the cost structure of his circuitry would be best served by making semiconductors from a single material. Unlike Kroemer's situation at Varian, Kilby's device was developed further by Texas Instruments after the air force committed to providing substantial funding.

The other industrial researcher who did not work for Bell Labs was Charles Kao at Standard Telecommunications Laboratories. Like Kroemer, Kao published his paper and launched his proposal to a skeptical audience. Although he sought industrial support, Bell Labs and others were not interested. However, he secured funding through the military and the post office, largely due to luck and timing. American management consultants had told the post office to spend more money on research, which allowed for the luxury of a small investment in fiber-optics and Kao's idea.

Industrial Research and Solid State Physics

The discovery of quantum mechanics was vital to inventions operating at quantum scales, from the transistor to the laser to magnetic resonance imaging. However, perhaps one of the most signature effects, explained by the Heisenberg uncertainty principle and wave-particle duality, is quantum tunneling. In quantum mechanics, particles can tunnel through potential energy barriers because they can behave like waves. A significant discovery in its own right, tunneling has also been important to inventions, including the scanning tunneling microscope (STM), and nanotechnology.

Although tunneling in the nucleus had been observed in the first half of the twentieth century, its potential and implications was not fully realized until observations were conducted at larger scales. All three recipients of the 1973 Nobel Prize in physics began their prize-winning work discovering effects of tunneling in various solids as graduate students. Their discoveries helped give rise to solid state physics and altered the fields of materials science and engineering. Perhaps it should not come as a surprise, then, that industrial laboratories were at the forefront of these discoveries and inventions.

Before graduating with their PhDs, Leo Esaki and Ivar Giaever worked in industry and invented experimental devices that led to discoveries about tunneling in semiconductors and superconductors, respectively, for which they shared half of the 1973 Nobel Prize. Esaki chose to work at Sony Corporation because of a desire to contribute to rebuilding post-war Japanese industry, though the year after he made this discovery, he left Sony for IBM. As an IBM fellow, he had the freedom to pursue his ideas, most especially the idea to make man-made quantum structures. As a materials scientist, he was interested in designing and engineering man-made materials. To do so, he needed freedom to exercise his creativity in both discovery and invention, and organized a group to help fulfill his creative vision. IBM made that happen.

Giaever joined General Electric's engineering department trained as a mechanical engineer. A few years later, after moving to their research laboratory, he began a PhD in physics. Despite the description on his diploma, he continued to think of himself as an engineer who needed physics because the mathematics applied to physical systems was more advanced than our understanding of the systems themselves. For Giaever, the large laboratory environment of GE and the resources available to him there were indispensable. As an engineer, he found physicists to be especially invaluable because he could take advantage of their expertise, materials, and experimental designs. While it was important that GE granted him the opportunity and freedom to move from engineering to research, Giaever also credits his immediate manager at the research laboratory, Roland Schmitt, for allowing him to pursue the physics of tunneling in superconductors instead of assigning the topic to the group's trained physicists. The other half of the 1973 Nobel went to Brian D. Josephson for predicting the behavior of tunneling for supercurrents. A graduate student at Cambridge at the time, his theoretical discoveries addressed questions brought up by Giaever's discoveries about how to calculate and predict tunneling currents.

The culture and community at IBM Zürich were just as crucial for the inventors of the Scanning Tunneling Microscope as they were to Giaever at GE. Gerd Binnig and Heinrich Rohrer, self-identified physicists, received one-half of the 1986 Nobel Prize in physics for inventing the STM. The STM provides clear, three-dimensional images of atomic structures by measuring changes in current due to tunneling. Binnig and Rohrer acknowledged the support from the scientific communities at IBM in addition to the management, which permitted experimentation, exploration, and mistakes. The innovation culture of IBM emphasized producing results, even if they might not be commercially useful to the company. In this case, the STM was first-rate research *and* useful to the company, so IBM pursued patent protection and continued work in the field.

Rohrer believed that this freedom to make mistakes was vital to innovation. The IBM research environment extended this freedom to physicists, who could pursue their own, independent projects. However, this same freedom was not extended to people working directly on technology creation. Although they were given the freedom to choose their own approach to solving problems, they had to work on defined projects within specified frameworks. While Rohrer distinguished between researchers and those working on technology, he considered the differences minor. Engineers, he believed, were also scientists. Rohrer found it best when scientists and engineers whose work overlapped worked together.

The other half of the prize went to Ernst Ruska for his discoveries in electron optics and invention of the first electron microscope, which for the first time allowed the human eye to behold the atomic scale. After inventing the electron microscope as a graduate student, Ruska returned to the source of his practical training—industry—to work on electron optics.

From Magnetic Molecular Rays to Nuclear Magnetic Resonance

Another highly influential set of inventions and discoveries can be traced in the family of Nobel Prizes given for nuclear magnetic resonance (NMR). These prizes cut across physics, chemistry, and medicine, providing a rich wealth of material for analysis. The NMR family begins with the 1943 Nobel physics prize, which was awarded to Otto Stern for "his contribution to the development of the molecular ray method and his discovery of the magnetic moment of the proton." Stern, adopting a method that had previously been applied to larger macroscopic phenomena, showed that it was possible to measure the magnetic moment of the proton. This method was superior to the existing spectroscopic methods, utilizing the deflection of protons in inhomogeneous magnetic fields to estimate the size of their magnetic moments. This allowed Stern to calculate the magnetic moment of the proton, and he discovered that it was 2.5 times larger than the theories had predicted. This achievement was predicated on his invention of the molecular ray deflection method. As such, in Figure 5.2, we describe this Nobel Prize as being firmly at the intersection of invention and discovery.

The 1944 prize given to Isidor Rabi built directly upon Stern's work, and had been recognized the prior year. This prize acknowledged Rabi's contribution to measuring magnetic moments. The challenge of Stern's deflection method was its dependence on the ability of the researcher to measure the surrounding inhomogeneous magnetic field at very small dimensions.

Rabi's insight was to use a different measurement entirely. Instead of measuring inhomogeneous magnetic fields, he would instead use the phenomenon of resonance to measure precessional frequencies. He realized that the interactions of the magnetic moment of particle rays with the surrounding magnetic field proceeding through precession movements around the direction of the field could be detected with the aid of a current carrying wire inserted into the magnetic field. The trick was that resonance occurred when the frequency of the electric current was the same as the precessional frequency of the magnetic moment of the proton. This led to much more precise measurements of magnetic moments.

This new invention, named the resonance method, linked work in radio frequencies with subatomic particle theory. The 1944 prize can thus be categorized as a new invention that built directly upon the discoveries recognized by the 1943 prize and opened up new avenues for very precise measurements of extremely small magnetic moments. In Figure 5.2, this prize is situated in the Invention half-circle.

The 1952 physics Nobel was given to Felix Bloch and Edward Purcell for their invention of new methods of nuclear magnetic precision measurements and the associated discoveries associated with their methods. The idea of nuclear induction is similar to that of electrical induction. In the presence of a homogenous magnetic field, the magnetic moments possessed by subatomic particles precess about the homogenous field, and a change in the homogenous magnetic field leads to a corresponding reorientation of the nuclear (magnetic) moment. This process can be compared to that of electromagnetic induction, where changes in electrical fields induce a current in an associated coil in the presence of the magnetic field. Bloch and Purcell showed how electromagnetic induction holds for nuclear magnetic moments and under conditions where weak oscillating fields are superimposed on strong constant fields, resonance conditions are met at the point when the frequency of the oscillating field is equal to the frequency of the precessional motion such that when nuclear reorientation occurs (described as relaxation), it induces a voltage difference in an external connected circuit. This brilliant insight allows the use of standard radio techniques to "see" and measure changes in magnetic moment, leading to fundamental insights about the structure of matter. This work is best described as a great invention that allowed for rapid strides in the discovery of fundamental insights about nature. In Figure 5.2, we place this prize at the intersection of Invention and Discovery similar to the case of the laser.

At the start of his acceptance lecture for the 1991 Nobel Prize in chemistry, Richard Ernst acknowledged magnetic resonance's debt to physics while also highlighting its utility for chemists, biologists, and medical doctors. Awarded the prize for his work in methodological developments in nuclear resonance spectroscopy, Ernst's contribution to NMR was in the area of pulse methods and Fourier transforms. Ernst addressed the major challenge of using NMR as a spectroscopic method, namely its sensitivity. At Varian Associates, Ernst developed a process utilizing pulse and Fourier transforms in NMR to sum multiple readings in a way that increased the signal-to-noise ratio and greatly improved the sensitivity of NMR. Ernst also filed for and received multiple patents for his work at Varian. After moving to ETH Zürich, Ernst and his research group made it possible to take readings in multiple dimensions by varying the pulse length. This work led to two-dimensional and three-dimensional readings, greatly enhancing the usefulness of NMR to chemical spectroscopy. In Figure 5.2, we have classified this as a methodological discovery and an invention.

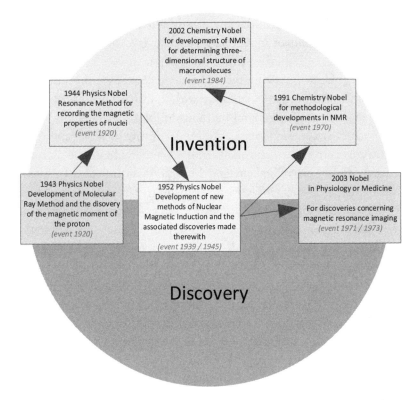

FIGURE 5.2. The innovation cycle in nuclear magnetic resonance. *Source:* Authors.

One-half of the 2002 Nobel Prize in chemistry was given to Kurt Wüthrich, Ernst's colleague at ETH-Z. Like Ernst, Wüthrich began working with NMR in industry after receiving his PhD. At Bell Labs, he began inventing a way to utilize NMR to determine the three dimensional structure of protein molecules, which he continued to pursue at ETH-Z. Using certain effects of dipolar interactions between different nuclei, Wüthrich was able to integrate this data with the information of the individual nuclei and build three-dimensional modules of biological macromolecules in solution, enabling interesting new studies of proteins *in situ.* This new method is properly classified as an invention, as shown in Figure 5.2, and is a further refinement of the previous work on NMR applied to the specific needs of chemical spectroscopy.

The final prize in our case is the 2003 Nobel Prize in medicine. Peter Mansfield and Paul Lauterbur shared the prize. Building on the previous work on NMR, Mansfield and Lauterbur both contributed to the development of magnetic resonance as a medical diagnostic tool. A chemist, Lauterbur began working with NMR while in the army and at Dow. After completing his degree and entering academia, Lauterbur continued to collaborate with industry. During that time, he began considering general methods to locate NMR signals on a nonuniform magnetic field. Lauterbur's contribution was the realization that introducing gradients into a magnetic field was useful in creating two-dimensional images. Mansfied continued in this vein, improving on Lauterbur's work and speeding up the imaging process by using mathematical techniques to reduce the time necessary to generate the images. The work of Lauterbur and Mansfied led to the invention of a new medical diagnostic device that allowed for unprecedented imaging capabilities and created a whole new modality of medical diagnostics, magnetic resonance imaging (MRI). Similar to the import of computer assisted tomography (which was rewarded with the 1979 Nobel Prize in medicine), NMR machines have transformed medicine. In Figure 5.2, we have situated this Nobel in the invention half of the cycle.

NMR Researchers in Academia and Industry

While many of the researchers mentioned in this chapter had ties to industry and the great industrial lab complexes, the 1952 Nobel Prize in physics is a fascinating outlier. After all, the 1952 physics Nobel was awarded to two physicists working in universities on opposite coasts of the United States. Because both recipients conducted their research in similar environments in the same field, it is not difficult to presume that they shared a similar potential for commercialization, patents, and profits. However, this was not the case.

Edward Purcell had a brilliant career in academia, disrupted only by working in the MIT Radiation Laboratory during World War II. His renown as a researcher and author of the classic textbook *Electricity and Magnetism* was not matched by financial gain; the book was published on a royalty-free basis in 1970, a mere five years after initial publication. Purcell's fellow prize recipient, Felix Bloch, also remained in a university setting at Stanford, but derived more profit from his efforts as a holder of patents and stocks related to his work. He was also skilled at placing many of his students as staff managers in industry. Bloch had a particularly strong relationship with Varian Associates that built on the Varian brothers' history and relationship with Stanford.

The contracts and arrangements between the Sperry Gyroscope Company, Russell and Sigurd Varian, and Stanford's physics department for patenting and developing the klystron before, during, and after WWII set a precedent at Stanford. The university developed a patent policy protecting and supporting the rights of its faculty, staff, and students in an environment rich in industrial connections, particularly with the founding of the Stanford Industrial Park, one of the original roots of Silicon Valley. Within this context, Bloch approached Stanford officials, including its president, about the university funding and patenting NMR. When they declined, Bloch then sought funding through the Research Corporation and naval contracts instead; the story could have ended there in a parallel to that of Purcell.

Russell Varian, who had convinced Bloch to apply for a patent with Stanford, offered his assistance with patenting in exchange for exclusive licensing rights to be granted the company, Varian Associates,which he and his brother Sigurd were establishing. The NMR patent proved to be profitable for both Bloch and Varian. Before Varian purchased the royalties for $1,571,000 in 1977, Bloch is estimated to have made many hundreds of thousands of dollars from magnet sales and several million if spectrometers are included. Though there were conflicts about rights of return and royalties based on the original agreement, the association remained fruitful. Bloch served as a paid consultant, stockholder, and resource for Varian.

Just as the NMR family of Nobel Prizes extends beyond Bloch's and Purcell's invention of techniques to measure the magnetic moment and the discovery of NMR, so too do the industrial connections for the subsequent discoveries and inventions. In fact, the next major breakthrough increased the sensitivity of NMR and formed the basis of modern NMR spectroscopy. This time, the work resulted in the 1991 Nobel Prize in chemistry for Richard Ernst, who conducted the early work at none other than Varian Associates.

After he finished his PhD at the Swiss Federal Institute of Technology in Zurich (ETH-Z), Ernst was interested in leaving the university environment because he felt "like an artist balancing on a high rope without any spectators." He sought additional motivation for his work by joining an American industrial lab with clear commercial goals. Because Varian employed famous scientists in the field, he decided to join them; he later credited his "incredibly brilliant coworkers" for his success. However, after inventing a spectrometer using Fourier transformations, Varian patented it but did not proceed to commercialization. Instead, five years passed before Tony Keller at Bruker Analytische Messtechnik demonstrated the first commercial Fourier Transform NMR.

The Discovery and Invention Cycle in Practice

The various prizes discussed earlier were drawn from fields as diverse as physics and electrical engineering, chemistry, biology and, physiology and medicine. This is a broad sample but by no means exhaustive. It is possible to reconstruct similar narratives from other fields of research such as computer science, various other parts of the chemical and biological sciences and other diverse fields of engineering which we do not explore here. From our studies though, a few critical points about the research practices and environments we have reviewed so far in this chapter are clearly evident.

For instance, contrary to expectations of anyone familiar with *Science, the Endless Frontier* the environment of the industrial laboratory was often a key ingredient in the dance between discovery and invention documented in our examples above. Rather than being inimical to discovery research, it appears that having a goal and a well-defined mission (as the industrial laboratories all did) catalyzes research that leads to both inventions *and* discoveries.[7]

Another point from our examples is the broad conception of research that operated at these institutions which allowed for a dual movement between building prototypes of devices, and also asking questions of a fundamental nature. The DIC examples above show how these two *research* activities are two sides of the same coin, and it is clear that radically innovative research flourished in environments that were supportive of this broad conception of research. Again, this refutes the problematic distinctions inherent in the "basic"/"applied" research framework.

In addition, from the previous DIC examples, there is a need for multiple kinds of expertise and high levels of interactivity between researchers. Physicists worked with electrical and mechanical engineers and also with metallurgists and other material scientists in collaborative and supportive environments where researchers were more interested in solving hard and challenging problems than in creating artificial boundaries.

This shift toward thinking about research in broader terms is also reflected in historical trends of Nobel Prize citations where we can see a clear acknowledgement by the Nobel committee of the cyclical nature of research as leading to both inventions and discoveries.

As an example, contrast the language of the citation of the 1956 physics prize for "researches on semiconductors and the discovery of the transistor effect" (rather than for the actual invention itself) with the citation for the 2000 Nobel chemistry prize to Alan Heeger, Alan MacDiarmid, and Hideki Shirakawa "for the discovery *and* development of conductive polymers." Both prizes were awarded to three individuals whose work was of a dual nature that resulted in invention and discovery. The transistor galvanized the field of silicon semiconductors and the work on conductive polymers galvanized developments in organic semiconductors and opto-electronic devices. Further examples of the Nobel committee's increasing acknowledgment of the importance of invention include, the 2009 Nobel physics prize to Boyle and Smith "for the invention of an imaging semiconductor circuit—the CCD sensor," and the 2014 Nobel Prize in physics "for the invention of efficient blue light-emitting diodes which has enabled bright and energy-saving white light sources" to Nakamura, Amano, and Akasaki. The historical record on the language of Nobel citations is illustrated in Figure 5.3.

A quick review of the language of the citations for prizes indicates that though there was a clear preference for giving prominence to discovery over invention, in the last few decades the acknowledgment of the inventive nature of research has been increasing in the citations of the Nobel Prize committee, a development which, in our opinion,

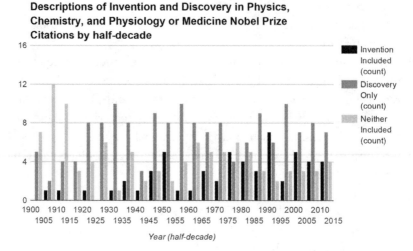

FIGURE 5.3. Trends in the Nobel Prize citations from 1900–2010. *Source:* Authors.

more accurately reflects the original intent of Alfred Nobel as indicated in his will. Interestingly this also correlates with the emergence of Nobel Prize winners from the industrial research laboratories.

In conclusion,

- First, all knowledge should be valued. Some knowledge production is oriented toward improving our understanding of the world through the process of *discovery*; some is focused on the creation of new techniques and devices, or on the synthesis and creation of engineered materials not found in nature, through the process of *invention*. The notion of the discovery-invention cycle attempts to view these two aspects of knowledge production as parts of a greater whole. Also by introducing new language, it allows us to escape the cognitive trap of thinking about research solely in terms of initial motivations, that is, "basic" and "applied." Rather, it urges the view of the research process as contextual and generative along particular lines of inquiry.

- Second, as demonstrated in the review of Nobel prizes above, we believe the issue of *research time horizons*—long vs. short— to be much more central to explaining radical innovation than theories about the applications and motivations of scientific research. Technological visions and ideas that focus on the long term are of crucial importance, and can lead to both novel science and novel technologies. Many of the industrial laboratories provided the right climate for the creation of future technologies and novel scientific discoveries.

- Third, we believe the discovery-invention cycle can assist in identifying *bottlenecks* in research. As it is essential to nurture every phase of innovation, this book should be read as an argument that we must ensure the adequate alignment of various institutions to foster the integration of ALL aspects of the innovation process (namely, invention and discovery).

- Finally, bringing together the notions of research time horizons and bottlenecks, we argue that successful radical innovation arises from knowledge *traveling the innovation cycle*. As just mentioned, all parts of the innovation process must be adequately encouraged for the cycle to function effectively, but the notion of traveling also emphasizes that we must have deep and sustained communication between scientists and engineers, and between theorists and experimentalists. Here, we move beyond a singular focus on motivation, to argue that we must bring all forms of research into deeper congress.

We have argued for a paradigm shift that ends the dichotomy of basic and applied research and bridges the divisions between science, engineering, and technology. This will happen if we embrace the more holistic integration of knowledge as it travels the discovery-invention cycle. Our studies suggest that such a rethinking is necessary for the design of more effective research institutions—where long time frames, a premium on futuristic ideas, and feedback between different elements of the research ecosystem are essential ingredients.

6

BELL LABS AND THE IMPORTANCE
OF INSTITUTIONAL CULTURE

AT THIS POINT IN the book it is necessary to reflect on the argument thus far. It should be clear that there are significant problems with the way science and technology research in the United States is prioritized and funded. It should also be clear that a large part of the challenge is the widespread embrace, particularly among policy makers, of the notions of "basic" and "applied" research. We have argued that there is a need to move beyond the problematic notions of "basic" research and "applied" research, and have proposed an alternative framework, the discovery-invention cycle.

This is all very well and good, and we would be delighted to see the widespread uptake of these ideas. However, even if notions of "basic" and "applied" are jettisoned, even if policy makers begin to treat research as an integrated whole comprising of both discovery and invention—such change would be insufficient without a number of complimentary actions. As was seen in Chapter 5, DIC research took place in research environments where the culture of the institution minimized differences between researchers, encouraged collaboration, and moved away from silos to a full embrace of research in all its various incarnations. It follows then that in addition to getting rid of the language of "basic" and "applied," to fully embrace the DIC, we also have to transition the cultures of our research institutions, moving them toward the characteristics described above. What are these cultural characteristics and how can they be instituted?

70

As we documented in Chapter 5, the DIC culture permeated the great industrial research laboratories such as General Electric, Bell Laboratories, IBM, and others. To understand this culture, this chapter analyzes the culture of the iconic Bell Telephone Laboratories, describing its various elements in detail.

The Bell Nobel Prizes

Ten of the sixteen Nobel prizes discussed so far in this book were given to people conducting research in industrial laboratories. Perhaps even more important, almost all the American laureates we have examined had industrial research experience, a large portion of which took place at Bell Laboratories. Of the six laureates working outside the industrial laboratory system, all were aware of the field defining research taking place in industry and saw the industrial laboratories as their premier competition. More than a decade earlier, Zhores Alferov and his Soviet colleagues were aware of the implications of heterostructures for semiconductor physics and technology, and by 1968 they were in a race with three industrial laboratories—Bell Laboratories, IBM Research, and RCA Laboratories.[1] We see that research, far from being constrained by association with industry, appears to have thrived and contributed greatly to both discovery and invention in the great industrial laboratories like Bell, IBM, and GE.

In *Science, the Endless Frontier*, Vannevar Bush famously argued that basic research needs to be protected from applied research. How then do we resolve the conundrum that industrial research, which ignored Bush's advice, was so successful that it was recognized with the scientific community's greatest honor, not once or twice, but multiple times? It is clear that the Nobel Prize winning industrial laboratories serve to undermine Bush's notions of applied and basic research. Somehow, in the industrial research environment, the very best discovery-oriented research was carried out. The point is that we need to pay attention to the connections between research and practice— the healthy connection between the two. It is our contention that a close attention to practices honed in industrial laboratories will reveal that the distinction that was most productive was not between basic and applied research, but rather between various research activities, and *product development*! That is, between unscheduled creative activity (both discovery and invention), and activities focused on achieving particular business ends (product development).[2]

This distinction can be clearly demonstrated through a close examination of Bell Laboratories, where the porous boundary drawn wasn't a question of methods, disciplines, or ability, but rather between research and development. The research area worked on unscheduled projects with long time horizons and no prescribed outcomes. Some of the time horizons were very long indeed (a decade or more). In fact, it was often difficult to tell when or if the work would turn into marketable products and that was fine. The development organization worked on projects with relatively well-defined specifications and shorter time horizons. Nevertheless, the staff on either side of this boundary had similar and complimentary training. Indeed, it was possible for a trained physicist or engineer to move from development to research and vice-versa.

Learning from Bell's successes, we suggest that there are two very important elements necessary to create and sustain the discovery-invention research cycle. The first is an acceptance of long time horizons for the solution of hard problems, plus a clear recognition of the unknown elements and unscheduled nature of research. The second is the institutional environment and culture that enables and supports good research practices.

There are a number of reasons for selecting Bell Labs for closer examination. For one, Bell Labs has the distinction of earning the highest number of Nobel Prizes of all the industrial research laboratories. During its heyday it was considered the top place to be for important research into solid state physics, materials science, electrical and systems engineering, computer science, and statistics. It attracted the cream of the crop from graduate schools all over the world. The combination of both the elite nature of Bell Labs, along with the distinguished trajectory of its many alumni, make it an important institution to examine. A large number of Bell Labs alumni have held important roles in academia, government service, and policy making. For example, Steven Chu as secretary of energy, Julie Phillips as chief technology officer at Sandia National Laboratories, Bob Dynes as president of the University of California system, Robert Birgeneau, chancellor of UC Berkeley, Cherry Murray as director of the office of science at the department of energy, to name just a few. As the birthplace of the transistor, the charge-coupled device (CCD), the UNIX operating system, radio astronomy, satellite technology and the LASER, and the discovery of novel phenomena such as the fractional quantum hall effect, laser atom trapping, and Cosmic Microwave background radiation, it is clear that a full understanding of the institutional culture that encouraged research of this caliber is of the greatest importance.

There have been many books written about Bell Labs and the various management styles individual leaders brought to the institution. One example is Jon Gertner's excellent examination of Bell Labs—*The Idea Factory*[3]—which provides an important glimpse into some of the accomplishments of the labs. Gertner's book is a useful journalistic review but it does not delve deeply into the structure of the institution that catalyzed the groundbreaking research. The other recent text on Bell Labs that *does* discuss some aspects of its institutional culture is the IEEE history center's project—*Bell Labs Memoirs*.[4] This collection of edited essays provides a number of important perspectives from individuals who worked at the labs and was a very useful resource in our thinking about the institutional culture of Bell Labs.

From its inception in 1925 Bell Labs was organized as a separate R&D entity, and through the next seventy-five years or so, it enjoyed an institutional culture that persisted through various management changes. William Baker, Mervin Kelly, and William Shockley were all prominent leaders at Bell who contributed in different ways to its culture of innovation. What were the elements that remained constant across different eras of management and contributed to the long and sustained run of innovative work at Bell? To understand this question, it is essential to examine the very start of Bell Labs.

A Brief Cultural History of Bell Labs

At the turn of the twentieth century AT&T was at a critical juncture. After the expiration of Alexander Graham Bell's initial telephone patents, AT&T had resorted to a strategy of protection to prevent competitors from gaining market share. By preventing other companies from connecting to its network, using lawsuits to hinder the competition whenever possible, and buying up equipment suppliers, AT&T had developed a reputation for bullying and unethical tactics.[5] It was into this environment that Theodore Vail rose to become president of AT&T in 1907 and instituted a shift in AT&T's strategy and culture. As a number of authors have argued,[6] Vail's previous experiences as a telegraph operator, and later in the US Post Office, molded his view of government and monopoly. Vail positioned AT&T to become a monopoly by arguing successfully that telephone service was essential and that the United States would be best served by a single integrated system. This idea was codified in the expression "one policy, one system, universal service." Though initially developed to reform AT&T's public image, this expression came to exert tremendous influence on the imagination of top AT&T management and resulted in a focus on long-term thinking and planning that shaped AT&T's culture throughout the twentieth century.

To create a single system and provide universal service, AT&T needed to solve the problem of long distance communication. The main technical challenge was the need to amplify signals so that audible communication over long distances (for example, from New York to California) would be intelligible. The best technology available at the time resulted in garbled, unintelligible speech. To resolve this, Vail created a research arm in Western Electric (the subsidiary of AT&T responsible for manufacturing) that reported directly to him. This division was led by an Electrical engineer, J. J Carty, and funded by a small tax on AT&T's operating companies. It was insulated from short-term pressures, and assured financial stability; managerially, it reported directly to Vail. Frank Jewett, a good friend of Nobel Laureate Robert Millikan and a member of the newly formed research group, was tasked with solving the amplifier problem. Given that all previous strategies tried by AT&T engineers had failed, Jewett turned to Millikan at the University of Chicago, and asked him if any insights from his physics research could provide new directions for exploration. Jewett also requested that some of Millikan's PhD students apply their laboratory expertise to the challenge of building a usable long distance repeater. This request resulted in Harold

Arnold joining AT&T and working with Jewett in the New York office. Arnold successfully refined Lee De Forest's "Audion"[7] device into a workable amplifier which we know today as the vacuum tube. The vacuum tube made possible the first coast-to-coast telephone call between Alexander Graham Bell and Mr. Watson.[8] Jewett's early success in recruiting Millikan's students led to the recruitment of even more of them, including Harvey Fletcher and Mervin Kelly. These new physicists worked under Jewett at Western Electric and made up the new department of research along with Harold Arnold.

After the war and the US Congress's exemption of telephony from antitrust laws, AT&T's domination of long distance service led to something of a natural monopoly. On January 1, 1925 the Board of AT&T created Bell Telephone Laboratories, Inc. in an effort to streamline the various research teams at AT&T centralizing the various research projects in the new entity. The new research and development organization was jointly owned by both AT&T and Western Electric. At the onset, Bell Labs was led by Frank Jewett as president and was funded jointly by AT&T and Western Electric through payments from the operating companies to AT&T under a license contract. Most of the thousands of Bell Labs employees worked on product development, but a few hundred worked on research. After the untimely passing of Harold Arnold, Jewett promoted Oliver Buckley to his position and appointed Mervin Kelly director of research in 1936.

At the tail end of the Great Depression, Kelly was in the unique position of having research jobs available at the Labs, and after the success of Clinton Davisson's research on thermionic emission from metals and the subsequent discovery of electron diffraction (which was honored with a physics Nobel Prize in 1937), Bell Labs became a highly sought after destination for newly minted PhDs from America's top universities. The lab's model of combining physical sciences and engineering continued on with the new hires. Among those that Kelly brought to Bell during that time were William Baker (joined 1939), John Pierce (joined Bell Labs in 1936), and William Shockley (joined Bell Labs in 1936), three people who would be highly influential on the Labs evolving culture.

Bell Labs allowed the research area to be insulated from the commercial pressures on AT&T and many of the signature elements of the labs, in particular the relationship between lab administration and the members of the technical staff, were put in place from the very start. The clearest statement of Bell Labs' culture is found in a lecture given by Ralph Bown, who was the vice president of research during the invention of the transistor. Speaking at the sixth annual conference on the administration of research at the Georgia Institute of Technology in 1952, his talk was titled "Vitality of a Research Institution and How to Maintain It."[9]

Bown described the nature of research that occurred at Bell as "activities, while not without concrete practical aims, are nevertheless not narrowly programmed to produce a particular result for a particular delivery date as their major purpose for being." Bown described this kind of research as unscheduled creative research, distinguishing it from development and design. He went on to identify a number of crucial elements that are necessary to maintain research vitality. First is the understanding that research environments reflect human relationships and group spirit. In short, successful research institutions should never forget that they are human institutions and they should place people above structure. Second, he noted like many after him (Chu,[10] Shockley[11]) the importance of a "well defined but broad technical objective" that allows for coordination and long-term planning. For Bell, of course, this broad objective was the advancement of communication technology. Here, Bown noted the importance of a common goal and defined objective in maintaining research vitality. Third, Bown discussed the essential importance of a robust but flexible organizational structure that, attentive to freedoms and dignity, allowed for a variety of expression by individual researchers. Finally, he discussed the importance of governance structures and the need for sufficient resources. Bown's most critical contribution though, was his direct revocation of the standard categories of research. "In my experience the desired understanding is not achieved by prefixing to research such modifying adjectives as basic, fundamental, foundation, applied, product, or any one of the other various qualifying names in common usage. . . ."

The success of Bell Labs and the evolution of its culture must be understood in relation to its position within the larger Bell System. The Bell System at its zenith was a massive and complex social and technical network that employed millions and touched the lives of virtually every community in the United States. As it transitioned into a fully regulated monopoly after the war, it increasingly began to see itself and be seen as a national resource.[12] The notion of natural monopoly and the previous nationalization of the Bell System during the war only served to reinforce the idea of telephony as an essential service and the Bell System as the best way to achieve full national coverage. This was encapsulated in the ideal of universal service. Indeed, the evolution of the Bell System took place against a backdrop of important legislation that served to shape the contours of telecommunications in the United States. One could even argue that the Bell System and antitrust and telecommunications co-evolved. It began with the Kingsbury Commitment in 1913 that saw Bell separated from Western Union and required Bell to open up its long distance network to smaller independent operators in an out of court settlement to an antitrust challenge. The rules were constantly being rewritten and tinkered with as the country struggled to determine what kind of communications system it wanted. The Kingsbury Commitment was replaced by the Willis-Graham Act of 1921 that allowed the Bell System to continue its policy of purchasing smaller local operators outright, enshrining the idea of the telephone system as a natural monopoly in law and helping to achieve Vail's notion of one system, one policy, universal service. It is against this backdrop that certain elements of the Bell System emerge.

We must remember that the entire argument for AT&T's natural monopoly status was based on the idea that one, national service was the best way to achieve high quality communication and interconnection. After all, in an openly competitive environment, there could be multiple systems all incapable of talking to each other and all suffering from poor voice quality (a scenario that seemed plausible after the initial patents expired and thousands of new, small telephone operators began competing with Bell). One of the reasons Bell received an exception for its monopoly status was so that it would maintain the quality of service for which it had become renowned. This expectation to deliver very high levels of quality of service had a number of important implications for the culture at Bell and ultimately the culture of Bell Labs. First, maintaining a high quality of service meant exerting and retaining control over virtually all aspects of the system's design. As a result, Bell (through the various operating companies) leased the phones, made exclusively by the Bell-owned Western Electric, in users' homes and charged a monthly lease fee. These devices were built to rigorous standards. One consequence of this was a sense of stability and a steady stream of finances. The requirements of the physical infrastructure and the interconnectedness of the system also meant that changes to any aspect of the system needed to be carefully considered. Any design changes had to be able to withstand the test of time as they would be in service for extended periods of time. These combined elements—the sense of stability from monopoly, the steady stream of financial resources, and the high level of system interconnectivity—meant that the entire Bell System operated on an assumption that designs and changes needed to be considered extremely carefully, with an eye toward long-term reliability. This long-term vision also helps explain why the administration was comfortable investing in a research department. The model of American innovation before the birth of the industrial research laboratory had been built on the work of the singular inventor (consider Edison, Ford, and even Bell himself). Even the "Audion" (vacuum tube), which was the very first success of Bell Lab's research division was a refinement of a device designed and purchased from a solo inventor (Lee de Forest). The success of integrating the processes of invention (epitomized by Bell himself) and deep scientific understanding (represented by personages of Arnold and Davisson) was the catalyst that led to the formation of Bell Labs.

In summary, the DNA of Bell Laboratories was a byproduct of AT&T's unique monopolistic position which delivered certain character traits that carried over into Bell Labs: namely, a concern with excellence, an orientation toward the long term, and a need for carefulness and thorough testing before implementing any changes.

These qualities are clearly seen in Mervin Kelly's own account of the evolution of Bell Laboratories after the invention of the transistor. In the *Bell Telephone Magazine*, five years after the invention of the transistor, Kelly reviews the early history of the transistor and evolution of transistor research at Bell over the previous five years.[13] He begins by reiterating the centrality of teamwork to the invention of the transistor and the discovery of the transistor effect. Kelly discuses successive generations of the transistor: from the first transistor; to the point-contact transistor; to the photo transistor; to the junction transistor, and then finally the tetrode Transistor. In his telling of the evolution of transistor research over five years, Kelly constantly oscillates between the interesting research and an awareness of how the transistor would fit into AT&T's broader mission of providing telephone service. Though it would take many years of testing and systems development for transistor-based phone system elements to become commonplace, Kelly's account is always sensitive to the company's broader mission. Along the way, he describes the many and varied contributions to transistor research of other researchers, metallurgists, and engineers. The link between research and development is clearly seen in the evolution of the transistor from an idea hatched at the laboratory bench into service in the broader Bell System. Kelly described the entire process as "organized creative technology."

In many ways what Kelly described as "organized creative technology" resonates with our notion of the discovery invention cycle and the idea of traveling the cycle. It stresses the importance and continuity of research and fundamental development, and the importance of maintaining the difference between the two. Kelly describes how shortly after the transistor was invented/discovered, Bell Labs formed a fundamental development group led by J. A. Morton, tasked with the responsibility of generating new knowledge of how the transistor could be designed and reliably replicated. According to Kelly, from the onset, the fundamental development team did work that was indistinguishable from that done by the research team, but as soon as the former had acquired the necessary knowledge, they continued to refine transistor design with a focus on generating designs that could be passed along to system designers for their use. This fundamental development fed back into the research process and involved the skills of crystal growers and metallurgists and led to the creation of new techniques and methods (for example, new methods for purifying silicon and making composite crystals). It was only after the completion of these processes that the baton was passed on to the manufacturing arm at Western Electric.

We should pause here to reflect on the elements of the Bell Labs culture that we have identified so far. It is clear that attending to the long term was carried over from the Bell System's DNA. Research is an unscheduled creative endeavor that by definition requires a long-term view. As the demands and time scale of industrial research fit perfectly with the needs of the Bell System it was uniquely positioned to innovate by integrating research capacity directly into its organizational structure.

Also, the institutional arrangements of Bell Laboratories within the AT&T system and the vision of its leadership allowed individual researchers the freedom to consider and investigate future-oriented technologies and to concentrate on cutting edge ideas, leading in no small measure to Kelly's description of Bell Labs as an institute of creative technology and also, in our opinion, a citadel of science. Bringing together technological futurists and researchers to work jointly on projects, all the while sharing a joy for the pursuit of scientific understanding, made Bell Labs the unique place that it was.

Finally, the unquestioned success of the vacuum tube reinforced the importance of recruiting the best and the brightest scientists and engineers from the nation's, and eventually the world's top universities to work in research for the ultimate good of the Bell System and the early leaders' experiences as academics meant that much of Bell's culture was modeled on university environments with all the freedoms and collaborations this implied. In addition, maintaining close contacts with leading university faculty was an important recruiting tool for identifying the most promising and creative PhDs.

Having considered cultural practices from an institutional perspective, let us now turn to individual perspectives and experiences at Bell Labs, continuing our exploration of Bell Labs' culture from the laboratory floor.

Elements of Research Culture at Bell Labs

To understand the culture of Bell Laboratories from the perspective of individuals who worked there required us to draw upon a diverse data set. Resources that reflect on the daily experience of the staff of Bell Laboratories are few. So we conducted extensive interviews with various Bell Labs research employees. The main source of data for the analysis below is our interviews.[14] In addition, we drew upon the recent work of the IEEE History center and the manuscript edited by Michael Noll and Michael Geselowitz—*Bell Labs Memoirs: Voices of Innovation.*[15] That project, which was designed to memorialize Dr. William Baker, is a collection of personal reflections on Bell Labs culture, especially during Dr. Baker's tenure. It is amazingly representative, covering various positions within Bell Labs, from research scientists to chauffeurs, and for our purpose it proved to be an invaluable resource. Finally, we relied on conversations about Bell Labs from a conference organized at Sandia National Laboratories in June 2013.[16] After reviewing all of our data, we identified seven critical elements of research culture at Bell Laboratories.

Freedom to Fail and the Patience to Succeed

Perhaps the greatest distinction between the research orientation at Bell Labs and research cultures elsewhere was the freedom to fail. Members of the research staff enjoyed this freedom in its totality. Simply, the freedom to fail is the idea that researchers are empowered to define their own projects even when they have no clear short-term utility. This freedom is a direct consequence of the nature of research, which is meant to be unscheduled and creative. The "freedom" implies that the institution supports individual researchers in their activities and sometimes provides broad guidance, but leaves the decision on what area to investigate to individual researchers. In *Bell Labs Memoirs*, John Pierce describes this freedom in this way : *"I received little guidance except that I was to do research on vacuum tubes.*[17]*"* Several of our interviewees experienced this kind of freedom from their very first day: they were given resources and laboratories and only the most general of suggestions. One interviewee responded by saying *"I walked into an empty lab and was told, 'Order what you need and do something interesting.'"* In other words, failure or success was in the control of individual researchers. The determination of whether or not a research program was successful was made by research administration, but the prosecution of the research program was up to the individual researcher, with resources and support provided by the institution. This freedom was only made available to research staff members (as distinct from those who worked in development) of Bell Laboratories. Perhaps the most important part of this freedom to fail was that *a project's failure was not necessarily counted as failure of the researcher.* This meant that individual researchers were energized to undertake lines of research without being worried about negative repercussions if they hit a dead end. This wasn't an open check to do poor research, but rather an endorsement of the individual researchers and their intellectual judgment. In contrast, the development teams experienced a lot more structure. This freedom to fail contributed to the research culture at Bell Labs in a number of important ways, least of which was a pervasive sense of empowerment among the research staff.

This also meant a lot of patience on the part of the institution with projects that were initially failures, but promised extraordinary returns if proven successful. As one of our interviewee's said, "they were given the freedom and the money and encouragement to work on that for six years. You won't find that today. Al Cho, who developed molecular beam epitaxy, I think he worked over ten years before he grew good crystals. Again, that was expensive." The freedom to fail sometimes meant the patience necessary to succeed.

Collaboration as the Primary Mode of Interaction

Bell Labs was an intensely collaborative environment. Every single one of our interviewees took this position. One of them told us that "Everything I did at Bell Labs was collaborative. I don't think I published a single paper on my own." The primary driver of collaboration was the kinds of problems that the researchers undertook to solve. Most of these research problems required multiple kinds of expertise. Collaboration at Bell Labs was about what each party could contribute to the research project in question. As one of our interviewees said, *"I don't think research collaborations are ever altruistic. Both sides have to get something."* The intense collaboration was necessary for producing the kind of field-defining work that Bell Labs is famous for, but these were not imposed collaborations. The collaborations were strategic. Individual researchers sought out others who possessed the necessary skills that were required to solve often mutual or overlapping problems. Each member of the collaboration usually got something out of it, perhaps a paper or two. Researchers worked in small groups, with at most a postdoc and a technician. This made it nearly impossible to either build an empire or to work alone. The size of the groups made collaboration the primary mode of interaction.

The interesting thing about the Bell Labs environment was that this kind of collaboration was considered a requirement of being a member of the research area at the Labs. Irrespective of how distinguished the Members of Technical Staff (MTS) became, and a number of them had won Nobel Prizes, all our interviewees testified to the open door policy and the expectation of intellectual engagement that permeated the Labs. The established norm was one of strategic and engaged collaboration and the institutional and administrative structures were organized to promote and to enable this collaboration. This norm of collaborative interaction was also reflected in the yearly evaluation process, allowing individual credit to be given to every member of a collaboration by each team leader. As an interviewee put it, "and the part of the culture that I found really helpful was that somehow it was ingrained in people that it was actually part of your job to cooperate, answer questions for other people so that if someone needed your expertise on something they were working on, that it was considered kind of part of your job to do that."

Competition as the Primary Mode of Individual Aspiration

Bell Labs was also an intensely competitive environment. The clearest example of how this was fostered was in the yearly review process. In the early days, every single member of the labs was ranked in order to identify the top 10 percent and the bottom 10 percent. As one of our interviewees said about appearing on the bottom 10 percent "if you got on that list a couple of years, you were kind of in big trouble and usually ended up working somewhere else. But I mean, it was a continuous process so every year you came up with it, there were always some people at the bottom no matter what was going on." And another said "everyone knew that it was kind of a competitive thing, survival of the fittest sort of thing. And you knew you had to face that, that you couldn't kind of slack off." This competition was partly driven by the extreme level of excellence represented by the research team at Bell. As an interviewee said "I could walk down the hall and find an expert in anything. I could find collaborators who would help me with anything. And that was about as good as it gets." Institutionally, there were occasionally multiple groups working on the same or similar problems albeit using different approaches. This redundancy generated a different kind of competition—competition among teams. A great achievement of Bell Labs was the fine balance it struck between competition and collaboration. One of our interviewee's captured it well: "I think you could think of it as a big family, a lot of sibling rivalry, a very successful big family. . . . There's a certain bonding that occurs that is really rather special. . . . I think it comes from that notion of it like being in a big family. There's some sibling rivalry, but everybody is doing fantastically well."

This competitive mode was clearly reflected in the high level of intellectual interactions that took place daily, and the expectation that one should always be ready to defend one's point of view. As Julie Phillips recounts, "scientists weren't allowed to rest on their reputations. The atmosphere was what I'd call irreverent and challenging. Bell Labs was not a polite or comfortable place. Technical debate was expected. If you were giving a talk or just engaging in a technical discussion around a coffee pot or in the hall, you expected that your ideas and data would be challenged. This forced you to think very carefully about your ideas and results before presenting them; you simply had to be prepared to defend them at great depth. All levels at the labs, up to and including the vice president of research, regularly engaged in these animated conversations. It was great fun—but it was not comfortable, and it was certainly not for everyone."[18]

Intense Interactivity with Peers

As a research environment characterized by competition and collaboration, Bell Labs was also a community that enjoyed intense interactivity. This was enabled by two processes. First, the infrastructure of the laboratory was such that offices and laboratories were located along long corridors (affectionately dubbed infinite corridors) and as Julie Phillips recounts, this resulted in "basically forcing everyone to walk along the same corridor to get from one point to another, made it inevitable that you would have multiple chance interactions with other scientists throughout the course of the day." Coupled with the norm of an open door policy, this resulted in multiple serendipitous interactions all day long. Also, the location of the Murray Hill facility meant that there were few to no restaurants within a walking, or casual driving distance. This led to a practice where most people would have lunch at the centrally located cafeteria. Not only would nearly everyone go to the cafeteria, but teams would go together *at the same time.* The collective team lunches meant that the likelihood of speaking to every member of one's team on a daily basis was extremely high. This sustained interactivity catalyzed research and helped produced the daily sharing of ideas. As Julie Phillips recounts,

> Nearly everyone went to the cafeteria at lunchtime, whether they bought food there or simply brought their lunch. The best tables were the large round ones, because it was always possible to squeeze in one more person. Sometimes the conversation was rather prosaic—politics, sports, and the antics of kids were familiar topics—but often the conversation turned to science. This was where the Bell Labs "stationery" (also known as a napkin) came in. It was often used to explore and then capture an idea that germinated during a conversation, which could then be tried out in the lab that same afternoon. Several of my publications resulted from just such interactions.

Administrative Leadership from Within

As has been discussed earlier in this chapter, the very first president of Bell Labs, Frank Jewett, was a physicist. John J. Carty, former vice president of AT&T and one time chairman of the board of Bell Telephone Laboratories, was an electrical engineer, and subsequent Bell labs leadership was drawn from physics, electrical engineering, and chemistry. The precedent established with Jewett as president was continued in the selection of administrators at Bell Labs. Anyone who made it into administration at Bell had to prove themselves successful researchers and prove that they had enjoyed accomplished research careers. In other words, they had to have proven themselves to everyone else at the labs before they would be considered. This was important for a number of reasons. One, the yearly evaluation process required each group leader to be capable of describing and defending the work of each member of his or her group. This meant that administrators at Bell Labs were technically proficient enough to understand the work that their group members were doing and were capable of explaining its importance. As one of our interviewee's said, "It wasn't simply a management kind of a thing. They certainly did that stuff, but they really did get involved at a real deep level and understanding what you were doing, why you were doing it, and why it was important. Well, sometimes they decided it wasn't important, but they decided that on a technical basis. I don't know if it's unique, but I thought it was really a strong point for Bell Labs."

This meant that team leaders could act as sounding boards and were able to discuss the fine details of the work being carried out. What was more, they could do this with anyone. "If you disagreed with someone, it didn't matter if it was your colleague down the hall who is a member of the technical staff or if it was [Bill Brinkman][19] or someone else and you had a discussion which sometimes looked more like an argument, and you really kind of go at it sometimes. And then I just can remember having some of those with Bill and then he would just stop and say, 'Yeah, you're right, that's what we should do.' Or I would say, 'Yeah, you're right, that's what we should do.'" This meant that there was a healthy scientific respect for team leaders by the members of their teams. It also meant that the administrators, having once been (and in most cases still continuing to be) active researchers, were oriented more toward serving the people on their team and acquiring the resources to make them successful. Another interviewee who became a team leader discussed this particular attitude of administrators at Bell "well, first of all, I would say that definition of being a manager in Bell Labs might have been a little different than somewhere else. It wasn't by any means telling people what they should do or organizing it. It was really more the other way around of providing them with what they needed." As another of our interviewees said, "The management was very much a part of it . . . they used to always say, 'You don't work for me. I work for you.'"

The tradition of administrators with a research orientation also meant a certain amount of flexibility and openness in the responsiveness of the administration. One of our interviewee's captured many of the threads we have mentioned above when relating an effort by the team she participated in arguing that credit should be allocated to the team, not to individual members. The team wanted credit allocated equally among all members. They were asking to not be evaluated as individuals but as a functional unit. "We were all, I guess, about of the same age. We weren't certain what success constituted. We were trying different things. We enjoyed being with each other and we enjoyed collaborating. And I think, there was a little bit of 'we have to educate our directors,' that this is important, and thankfully we were successful [in research efforts], because if we hadn't been successful, they wouldn't probably have listened to us. But what we were trying to do together with equal contributions by everyone was important, and we felt that we should be evaluated as a group. It wouldn't be like 'I did this and he did that.' The important thing was that our managers understood what we wanted. They were accepting, they took that into account, they defended us, they allowed us to work together, go forward as a team, and purchase new equipment. The fact is that young as we were, when we started at Bell, we were given full technical support, we had administrative staff to help implement and realize our ideas, and we had the necessary people. It was my first job out of graduate school. I think that as much as I realized that it was a wonderful place to work in, I didn't know what the other places were like. In a way, I was too young to appreciate it until I went away from Bell Labs and understood in other places, every time you make a Xerox copy, you have to have an account to charge it to."

The administration at Bell Labs consisted then of highly accomplished researchers who were flexible enough to allow for new kinds of collaborative organizations and were capable of understanding the technical and scientific merits of the work of their teams. Their entire orientation was to help, serve, and assist their group members to obtain the necessary resources to be as successful as they could be. Finally, the administration was also capable of discussing the technical merits of various projects and determining their value. These determinations were not always correct, and sometimes accomplished researchers didn't always make good administrators, but a large number of them did amazingly well and their success was in no small part responsible for the enabling research environment at Bell.

Egalitarian Meritocracy

Bell Labs was an egalitarian meritocracy—the idea that everyone in the labs was equally deserving of respect for their ideas and contributions was deeply ingrained. All of the people we interviewed expressed this idea in different ways. Perhaps the most common expression of this idea was the experience of the "impostor syndrome." As one of our interviewees (who is widely regarded as the foremost investigator in his field) recounted, "I kept thinking, they're going to find out . . . I'm going to get my pay slip, and you know, and I spent the first five years thinking, 'God, it is just a matter of time.' I don't measure up . . . then I transitioned to they're not going to fire me. They're just going to move me over to the development area, when they find out that I have been faking all along. But those same people would come to me with questions, how do we cut the molds and the laser? How do we do this? Can we do that? And they caused me to think, and sometimes I had a contribution in my own way." Many of our interviewees recounted a progression from arriving at the Labs, surrounded by "the people who wrote the book" in their area of interest and fearing that they would never fit in, only to find out that even Nobel Prize winners were interested in their work and their ideas. This near instantaneous acceptance *and engagement* by highly accomplished peers led to a quick sense of empowerment.

One example of meritocracy and egalitarianism at work is reflected in the research titles at Bell Labs. Until very recently, everyone involved in research was classified as a member of technical staff (MTS). This MTS designation was the only designation of the research staff, and was held by Nobel laureates and newly minted PhDs. Starting in the 1980s, superlatives like "distinguished" began to be appended to MTS, but for most of the lab's history, MTS was the main designation for all research staff. There were salary distinctions, of course, but the uniform title was an amazing leveler. Also, MTS wasn't only limited to PhDs. In rare and exceptional cases, others who had proven themselves capable could also become MTSes. For example, a number of technicians were so valuable, creative, and productive, that they became MTSes.

The sense of meritocracy also extended to the technicians in the laboratory. As researchers spent extensive amounts of time with their technicians, very close, near equal relationships would sometimes result. As one of our interviewee's recollected, "we're just the two of us in the lab, all day, every day, and it was a very close collaboration. My wife and I would go to his family picnics. He was at our wedding. . . . So, it was a very close relationship. No one had questions about hierarchy. It was much more of a partnership than a "Me Tarzan, You Jane" kind of thing. . . . At Texas Instruments, it was the opposite. It was very clear that technicians were a lower order. They were evaluated with very different principles. . . . It wasn't the case at Bell at all. In fact, some of the technicians became associate members of technical staff. And then, I think, maybe some of them actually became members of technical staff, without PhDs because they were so good. That would never have happened at Texas Instruments. So, that was the difference. And they also had, you know, my technician at Texas Instruments, I had to sign her time card. . . . My technician at Bell, you know, he decided that he needed to work on something else, that he needed to run that morning into New York City. He would call and say, hi, I can't come in today, and that was that. I would respond and say okay, no problem."

One upshot of the egalitarianism, was that technicians became cocreators. Again, the same interviewee comparing the relationship between technicians at Bell Labs and technicians at another industrial laboratory: "My technician at the previous industrial lab where I worked was very good at what she did. She could do diffusions, photolithography, metal contacts. However, she would never have suggested to me, 'You know, we could change this measurement, we could make this device better this way if we did this.'" She wasn't a part of the creative process. My technician at Bell Laboratories was part of the creative process. He was a, we can do this, in a new way kind of person . . . In fact, we made him in charge of doing life testing. He figured out how to do it. He set it up. He interpreted the data. It would never have happened at the other industrial research laboratory."

Excellence as a Virtue in Hiring, Promotion, and Review

From the early days of Bell Labs, when Frank Jewett turned to Millikan's research group as a hiring pool for Bell Labs, new hires at Bell Labs came from the nation's top universities with hires from Harvard, MIT, Stanford, Caltech, Columbia, Cornell, Berkeley, and Princeton among others. Hiring at Bell was taken very seriously and fully engaged the administrators. It was not uncommon for department heads, directors and executive directors to also be recruiters. As was the case from the early days of Frank Jewett's relationship with Millikan, recruitment at Bell focused on building relationships with the relevant research departments and faculty members at the targeted school. Recruiters visited every year, and relied greatly on the academic faculty to recommend particularly promising students who were then actively pursued. This relationship-based process began with campus visits followed by a prospective candidate receiving an invitation to visit in person. These visits and the requisite research presentations were legendary for their contentious and sometimes combative intellectual jousting. Candidates were expected to be able to handle any questions raised during their presentation and the pressure that came with such scrutiny. The candidates were the responsibility of the recruiter during their visit, but it was the intellectual engagement with other members of the labs that determined how the candidate's visit went. If a candidate wasn't the right fit, the recruiter could end up with a candidate on a multiple-day visit with no scheduled meetings. As one of the interviewee's recounted:

> "So, for me, I went through graduate school in Berkeley, and there I experienced the Bell Labs recruiting, which was very special, and the Bell Labs recruiters visited Berkeley every year and they followed up the progress of the students in Berkeley as they did at many, many other universities. So that was a very effective recruitment tool for the labs. It was also good for me.

My recruiter was Bill Boyle, who at that point was a recruiter and department head and subsequently won the Nobel Prize for the charge-coupled devices. So he lined up my visits at Bell Lab and that consists of two days of visiting both the research area where I was interested in seeing a postdoc and visiting many, many other areas."

Promotion and review at Bell Labs were done yearly, with a focus on quality not quantity. Individual researchers filled out single page self-assessments and then these were passed along to department heads who then came together and ranked everyone. The responsibility to promote and explain the work of individual researchers lay with the department heads who had to be very knowledgeable about the work of the people they led. This review process resulted in identifying people at the top, middle, and bottom of the pack. Using a roughly sliding five-year window, people at the bottom of the pack would feel pressure and in a few years be asked to move on unless their work improved. This constant pruning and categorizing resulted in a ubiquitous excellence. As one of our interviewees put it: "The average Bell Lab scientist was more often than not much better than people outside. It sounds arrogant, but I can tell you that that was the case." This yearly review wasn't about identifying people to move out of the labs, but was really focused on encouraging researchers to do their best work, even if sometimes that meant encouraging them to move into other areas. "People knew if you were not working anymore at the top of your abilities. Really, you couldn't hide. And that was good for the organization. Of course, it was tough. Quite often you would tell people . . . I remember as a manager too, I said, pushing them to change fields. Many resisted. Some resisted because it's not easy. You're working in a field because you carry the inertia of having become famous, but insiders know that your best days have passed because you're working on perfecting stuff that you have done before. But the people that got the message, some went on to a new great career and were actually grateful after they made the changes. . . . There are many stories like this. It was a culture of continuous performance."

The Interaction of Institutional
Structure and Research Principles

The various elements of the institutional culture at Bell Labs that we have discussed above can be seen clearly in the institutional *structure* of the laboratories. Figure 6.1 offers a diagrammatic representations of the Bell Labs institutional structure circa 1976.The figure shows the clear relationship between the principles we have discussed, and the institutional design of the laboratories. A significant portion of the material used in the following sections are derived from Narayanamurti's extensive experience at Bell Labs, both as a member of the technical staff in the Physical Research Laboratory[20], as a department head in the Solid State Electronics Research Laboratory and later as the director of the same laboratory.

As shown in Figure 6.1, Bell Labs was jointly owned by AT&T and Western Electric and was funded by a tax on AT&T Longlines and the Local Operating Companies (which in turn were also owned by AT&T). The lab was run by a president who had as two direct reports, the executive vice president of systems engineering and development, and the vice president of research and patents. The majority of research activity at Bell Labs happened in the research and patents area (called Area 10). It is a mark of the importance of the research area that the vice president of research and patents reported directly to the president and not to the executive vice president.

Operationally, the five other areas (Areas 20 through 60) reported to the executive vice president of systems engineering and development and were together about ten times as large as the research area. This relative size difference between research and systems engineering and development is sometimes forgotten in discussion of the labs and its accomplishments. In addition, it is important to note that the research department also took on the responsibility for patents, clearly pointing to the understood link between the activities of research and patenting at Bell Laboratories.

In Figure 6.2, we learn a bit more about the culture of Bell Labs research area in 1976. We can see that the vice president had six executive division directors as direct reports. Even the names of the various divisions give us insight. Notably, each division had the term

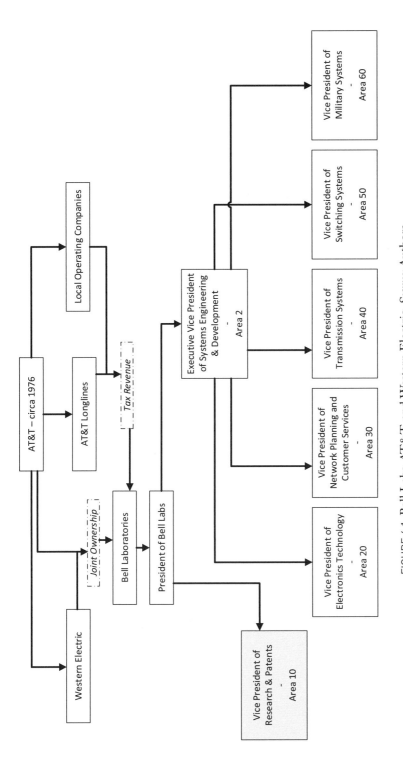

FIGURE 6.1. Bell Labs, AT&T, and Western Electric. *Source:* Authors.

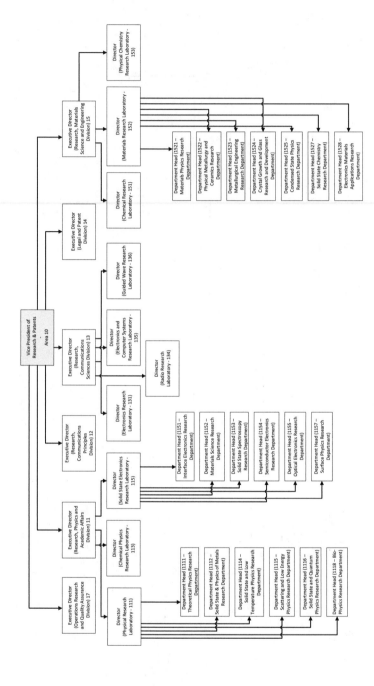

FIGURE 6.2. Bell Labs research structure in 1976.

"research" in its name. It is also important to point out that the legal and patent division was also led by an executive director.

One of the first things to stand out on close inspection of Figure 6.2 is the width and depth of the organization. The scope of the organization was wide enough that collectively, the laboratories provided the scale necessary to support research activities into all aspects of AT&T's telecommunications business. From the information technology contributions of the computer science research group (UNIX and the C language, for example), to observing new states of matter in the fractional quantum hall effect, Bell Labs had the scale to tackle all research areas pertinent to AT&T. On the other hand, the labs also had the necessary critical mass of researchers necessary to produce field defining work in key areas. This balance between critical mass and scale was enabled by the generous resources of AT&T. It is doubtful if any private company can afford to fund a similar institution; its closest parallel today would be the National Laboratories.

Due to limitations of space, not all the divisions are fully sketched out, but those that are do reveal important information. Take division 11, for example, the research, physics, and academic affairs division was widely acknowledged as the most "science oriented" division at Bell Labs research. Interestingly, however, academic affairs and recruiting wasn't relegated to a separate department but was a core function of division 11. The three laboratories in division 11 were the physical research laboratory, the chemical physics research laboratory and the solid state electronics research laboratory. A close examination of these individual laboratories reveals Bell Laboratories' deep commitment to interdisciplinary research.

The physical research laboratory included a bio-physics department that was perfectly comfortable hiring nonphysicists to work in it. The commitment to redundancy and competition can also be seen in the very similar work of the departments 1114 (solid state and low temperature physics) and 1154 (solid state spectroscopy research), and in the existence of a surface physics research department (1157) in the solid state electronics laboratory. This redundancy meant that these research teams sometimes conducted similar work and would even compete to hire the same person, reinforcing Bell Lab's culture of competition as the primary means of inspiring individual effort.

The same institutional design that lead to different research teams covering similar ground also encouraged collaboration (as resources could be shared among the various research groups. Division 15 was given over to engineering and materials science research. Finally, development was also brought into the research division. Department 1524 was given over to the research and development of crystal growth and glass fibers. The structure of Bell Labs research reflects the porous boundaries, the commitment to interdisciplinary work, and the intersection of science and engineering. As we have discussed, the value of egalitarian meritocracy found at the labs meant that distinctions such as "scientist" or "engineer" were effectively effaced in the daily life of the labs.

The distribution of Nobel Prizes demonstrates the commitment to excellence in all aspects of the research mission. The Nobel Prizes for the transistor and the laser came from the physical research laboratory (physics and academic affairs, division 11), Steven Chu's light trapping Nobel Prize came from the electronics research laboratory (in the communication sciences, division 13), the Nobel Prize for the fractional quantum Hall Effect was given for work done in the solid state electronics research laboratory (laboratory 115, division 11). The 1977 Nobel Prize in condensed matter physics given to Phillip Anderson emerged from the physical research laboratory, and the discovery of cosmic microwave background radiation (CMBR) that Arno Penzias and Robert Wilson were responsible for came from the radio research laboratory 134 (in the communication sciences, division 13). The work on CMBR won the 1978 physics Nobel Prize. The Nobel Prize for the charged-coupled device (CCD) came from Area 20, the area responsible for electronics technology development.

Rethinking Institutional Research Cultures

There is no doubt that the institutional culture at Bell labs has been highly influential. Its heyday is regularly invoked in discussions about the strength and resilience of American Industrial research. It is clear, however, that Bell Labs was a product of the convergence of a culture and technology that depended in no small part on AT&T's unique circumstances and market position. While we don't think that Bell

Labs should or could be recreated, there are elements of the Labs' culture that would improve the quality of research in any institution focused on similar long-term transformative research such as Bell Labs was created to undertake.

The seven elements of the unique culture at Bell that are explored above are transferrable in one way or another and are not dependent on the benefits of a monopolistic position in the marketplace. They are elements that every research institution can seek to emulate. Each of the seven represents investments for the institution and individuals employed by it. Institutional culture is a product of policy, infrastructure, norms, and leadership. Leadership is perhaps the most important element in the creation of institutional culture and the administrative leadership in any research institution holds the greatest responsibility for fostering the culture described above. Allowing for the freedom to fail and demonstrating the patience to succeed, promoting meritocracy, encouraging focused and engaged hiring to identify the appropriate fit of prospective individuals, and being engaged with and aware of research projects are things that administrations can cultivate. Furthermore, the other elements, including the fine mix of collaboration and competition that generates excellence at the individual and team levels, are elements that are influenced by institutional policy and infrastructure.

From our experience with various research cultures in academia, industrial research laboratories and government research labs, we surmise that a large portion of these elements are sorely lacking. The daily experience of academic researchers, with few exceptions, has become solitary with some colleagues in the same department conversing perhaps a few times a year. If it is so difficult to get sustained intellectual interactions with faculty members in the same department, how much more difficult must it be for colleagues with different kinds of expertise to engage in them? Some industrial research laboratories (such as Google and Microsoft) have done a good job of increasing interactivity, but there is no doubt that academia could benefit from efforts to improve, a point that will be explored in greater detail in Chapter 7.[21]

Another element that deserves extended attention is the freedom to fail. One of the amazing elements of the integrated research culture at Bell Labs was the amount of redundancy that was encouraged by the administration. It was not uncommon for different researchers and different research teams to work on the same problem, albeit from different perspectives. Internal competition was as important as external competition. This required a level of comfort with some research lines not being successful, and failure was not held against researchers. This encouraged ambitious research agendas and incentivized bold ideas. Our national research laboratories and funding institutions would benefit from embracing this lesson.

Finally, the integrative nature of research at Bell which was closely coupled with the broad mandate of AT&T made for a very delicate and finely tuned research/development organization. The important thing to remember is how catalytic research and development were to each other. Development (especially the fundamental development teams) obviously relied on the work of the research teams, but the challenges they faced often posed new and important questions that the research teams took up. Also, the barrier was not so inflexible that new ideas could not come from development. Indeed, as has been pointed out above, the work that led to the invention of the charge-coupled devices widely used in imaging applications was undertaken by Willard Boyle and George Smith, two Bell Labs' employees who were at the time working on the development side of Bell Labs.

7

DESIGNING RADICALLY INNOVATIVE
RESEARCH INSTITUTIONS

As we discussed in the previous chapter, the discovery-invention cycle requires new language and a specific cultural environment to thrive. This goes hand-in-hand with institutional structure. There have been various attempts to redesign institutions in such a way as to mitigate the worst elements of the "basic"/"applied" framework, and various blue-ribbon reports have called for new institutional structures and categories.[1] Our examination of Bell laboratories and the institutional structure of the labs in the previous chapter leads naturally to the question of if and how a similar institutional structure (along with its associated culture) could be implemented in contemporary institutions. Is it possible to design current day institutions on similar principles?

Undertaking an effort of this sort requires rethinking all the major elements of institutional design and culture—a colloquial expression from Bell Labs captures this perfectly—"You don't invent the transistor by continuing to improve the vacuum tube." This rethinking needs to include how the institution is managed, the structure of the institution, the structure of incentives, and even the institution's physical architecture. The good news is that there are successful and current examples of institutions that are designed along similar principles of the discovery- invention cycle and can serve as exemplars of what is possible when research is approached holistically rather than divisively.

This chapter will undertake case studies of two institutions that have actively worked to promote an integrated view of research and will attempt to identify and examine the various strategies that have worked to undermine the knowledge silos that prevailing conventions encourage. To showcase what is possible on a university campus, one case study is drawn from academia. The other case study highlights the increasing importance of nonprofit research funding, especially now that industrial funding is on the decline. Both organizations took different paths in escaping the basic/applied trap and generated unique solutions that were appropriate for their specific institutions. There really is no single path.

In addition, both of these institutions naturally embrace the discovery-invention cycle as it was modeled in Bell laboratories. In the one case, Janelia Research Campus was intentionally modeled in large part on Bell Laboratories by its founding architects, and in the other case, University of California Santa Barbara (UCSB), sometimes affectionately known as Bell Labs West, was inundated by Bell Laboratories members of technical staff who left Bell Laboratories at the time when it began shrinking after the breakup of AT&T. It is no wonder then that these two institutions embody the principles of the discovery-invention cycle in their institutional design.

It is our hope that these case studies will provide inspiration for possible strategies to move similar institutions away from the basic/applied dichotomy toward a more holistic view of the research process. The material presented in the UCSB case study is based on extensive interviews with various faculty members and former and present deans and vice chancellors at the UCSB conducted in the summer of 2014. The material in the Janelia Research Campus case is also based upon extensive interviews and visits to the campus.[2]

The University of California Santa Barbara

The University of California at Santa Barbara (UCSB) is one of the larger campuses in the University of California system. The University of California is a behemoth of a public university system with ten campuses, more than 15,000 teaching faculty, and over 230,000 students. Of the ten campuses, the Santa Barbara campus is of particular interest to our analysis as it exemplifies the institutional elements that concern us in this chapter, those that are essential to the creation and sustenance of a robust interdisciplinary environment especially in engineering and the physical sciences. UCSB has morphed from being a small independent teacher's college into a major world class research university in approximately three decades. As exemplified by the most recent National Research Council's evaluation of doctoral research programs, the university had close to half of its assessed doctoral programs in the top ten programs in the country. Ten of the programs were in the top five. These include, the materials department, which was ranked number one over the entire assessment range, chemical engineering, computer science, geography, marine science, mechanical engineering, communication, theater and dance, electrical and computer engineering, and physics. The process through which the university went from arguably humble beginnings to a distinguished research institution is the focus of this section of the book. The institution has become home to six Nobel laureates—beginning in 2000 and culminating in the 2014 prize in physics awarded to engineering professor Shuji Nakamura for his invention of the blue light emitting diode.

Our discussion will focus on the critical elements that led to UCSB's rise in the physical sciences and engineering. This can be traced to two important events. The first was the decision of the physics department to compete for an NSF-sponsored national center for theoretical physics, and the other was the establishment of the materials department in the College of Engineering. We begin with a discussion of the Kavli Institute of Theoretical Physics (KITP) for reasons that will become obvious, and then continue with a more detailed recounting of the transformation of the College of Engineering.

Kavli Institute of Theoretical Physics

The Kavli Institute of Theoretical Physics at UCSB was described in the *New York Times* as the "most important program for visiting physicists in the world," by Charles Munger of Berkshire Hathaway fame. Munger's comment was made on the occasion of his $65 million donation to the Institute in October of 2014—the largest single donation in the history of UCSB.[3] Continuously funded by the National Science Foundation since its inception in 1979, KITP began life as the Institute of Theoretical Physics (ITP), which was formed following a call for proposals for a "national laboratory for theoretical physics." This notion was championed by Boris Kayser at the NSF, and the UCSB physics department submitted a very compelling application. UCSB physicists proposed a unique institutional structure. They envisioned a research institute with a very broad mandate to engender new conversations about potential research programs through the interactions of visitors who would undertake the intellectual deliberations while physically in residence at UCSB. Notably, the visitors would be supplemented by a small number of permanent members of the Institute, who in turn would be tenured professors of physics at UCSB and would provide leadership for the Institute. At the time, this was a novel and exciting institutional structure and was designed to be open-minded and interdisciplinary. It is interesting to note that the interdisciplinary nature of the ITP was itself informed by the experiences of a number of its early permanent staff members, many of whom were influenced by the ideas of Sir Rudolf Peierls of Manhattan Project fame.

After Peierls' work on the atomic bomb, he left Los Alamos and returned to the University of Birmingham in the UK, where he established the Department of Mathematical Physics. The department instituted a robust visiting program for postdoctoral fellows and visiting students. The community that formed around Peierls in Birmingham was deeply influenced by his notions of the open-mindedness of physics, that is, the importance of physicists attending to everything that could conceivably be considered to be part of theoretical physics. This attentiveness to the widest possible range of exploration in theoretical physics was also championed by Nobel Laureate Walter Kohn, who became the first director of the ITP and who established much of the current interdisciplinary culture of the institute.

Since its inception, ITP has had six directors and has achieved global renown as one of the most important sites for the advancement of theoretical physics. After a large donation in 2002 by the business mogul Fred Kavli, the ITP was renamed the KITP. The institute's primary function in the theoretical physics community lies in its programming. It provides intellectual leadership by identifying emerging and important areas of exploration and brings together critical and influential voices to catalyze these new, interesting, relevant, and groundbreaking conversations. The residential nature of the institute and the close and sustained interactions of the participants creates an environment ripe for creative interactions that literally drive physics forward as a discipline.

KITP has been ranked number one in its impact on research conducted at major research universities and national laboratories.[4] It averages about 1,000 people per year participating in its programs and activities, and the average length of a visit is thirty-six days. KITP has about six permanent faculty members at any given time, and they, along with other affiliated researchers, constitute its core.

Establishing the ITP was a major accomplishment of the university and its then Chancellor Robert A. Huttenback. Though Huttenback's tenure at UCSB ended in controversy, there can be no doubt that his risk-taking paid off with UCSB's successful NSF bid in 1979. By providing the funding for the permanent members of the Institute from university resources, Huttenback made a large commitment that helped to persuade the NSF to fund the creation of the ITP. Though this led other parts of the university to complain, Huttenback's decision has been validated by the success of KITP. After successfully building up theoretical physics at the university, the chancellor decided to focus on building the engineering college.

Engineering at UCSB

A few years before the physics department submitted its proposal that led to the ITP, the engineering college at UCSB underwent a crisis of confidence. Ed Spear, the chair of electrical engineering, was hoping to shake up his fledging department. At that time, the college was still young, having only begun in the early 1960s. As luck would have it, Herbert Kroemer (who would go on to win the 2000 Nobel Prize in physics) was ready to leave the University of Colorado and had expressed an interest in moving to UCSB. Kroemer's initial interview with Spear was confrontational, aggressive and critical. In essence he thought poorly of the direction the department was headed in, and wanted to launch a brand new research program in compound semiconductors. At the time, silicon semiconductors were all the rage, and were the focus of most of the large research programs at other institutions. In short, Kroemer seemed to be a perfect fit for a Chair who wanted to shake things up and so he was hired in 1976. Shortly afterward, Kroemer was given the resources to hire along the lines he had indicated to Spear, and convinced Jim Merz, a leader in semiconductor research, to leave Bell Labs and help him build up a compound semiconductor research program at UCSB's College of Engineering. This new initiative in electrical engineering began just before chancellor Huttenback decided to commit new resources to build up the engineering college.

The Chancellor's commitment to engineering at UCSB began with the recruitment of a new dean of engineering, and the search committee (whose members included faculty members of the ITP) recommended Robert Mehrabian, who was hired in 1983. Mehrabian was hired away from the National Bureau of Standards where he ran a successful material science program. Mehrabian, by all accounts, was a very dynamic and forceful leader with clear ideas about the future direction of the college. As a condition of his hiring, he negotiated for and received a commitment from the university administration to allow him to fill fifteen new faculty positions and commensurate financial resources to create a new materials department in the College of Engineering. Of course, this allocation of resources would not have been possible without the full engagement of the Chancellor and his prior desire to allocate substantial resources to engineering at UCSB.

Mehrabian embraced a broad ideal for the new materials department reflected in its very name. The materials department, without the descriptor of "science" or "engineering." He also adopted Kroemer's notion that in order to distinguish the program, they would have to invest in emerging areas of research and avoid retreading the ground that the larger engineering schools such as Stanford and Berkeley had already covered. This meant pursuing compound semiconductors instead of silicon, for example. We should note that the study of materials by its very nature encompasses many aspects of the sciences such as physics and chemistry as well as many aspects of engineering. The decision was made by UCSB's leadership to focus on three distinct areas of materials: electronic-materials, with its strong relationship to electrical engineering; structural-materials, with its clear connections to mechanical engineering; and soft condensed matter, with its ties to chemical engineering. The program was thus envisioned as an interdisciplinary and interdepartmental effort with virtually all members of the new materials department also holding appointments in sister departments such as electrical, chemical, and mechanical engineering.

The new positions in electronics materials were filled predominantly in the area of compound semiconductors, with a majority of the faculty coming from industrial laboratories such as Bell Labs and Rockwell International Science Center. In these recruitments, Mehrabian, Kroemer, and Merz were aided by luck, timing, and external forces. Bell Labs, for example, was undergoing transitional pain and turnover as the era of the AT&T monopoly came to an end and as it became increasingly clear to researchers at Bell Labs that their storied institution would never be the same. In the area of structural materials—Mehrabian's area of expertise—the new dean succeeded in recruiting Anthony Evans, a preeminent researcher in the field, as the department's founding chair. Tony Evans recruited other distinguished researchers to the new department, often in partnership with the other departments at UCSB. Strength in areas of soft materials such as polymers and complex fluids came from the recruitment of industrial researchers from highly respected industrial research laboratories such as Bell Labs and Exxon Research (which was also undergoing restructuring at the time).

These various elements: Chancellor Huttenback's desire to build a great engineering college; Kroemer's laser-like focus on compound semiconductors; Tony Evan's stature in the field of materials; Mehrabian's tenacity and willingness to forge a new institutional path for the college; and finally the slow dissolution of Bell Labs created a perfect storm that allowed UCSB to hire away, or in Kroemer's words "raid" Bell Labs and other industrial laboratories, building strength in materials—almost overnight by academic timeframes.

Building Engineering at UCSB

The new materials department was envisioned as a research-focused department, accepting only graduate students, and the decision was made early on to use the opportunity to build a new culture of research within UCSB with the hope that over time it would permeate the campus more broadly. As Kroemer explained, the early team had three criteria in selecting their new hires:

> The first thing is how good is this person regardless of viewpoint? If the person is not first rate, forget it. I never would hire—make a substandard hire just for reasons of programmatic need. I would rather let the program stop. The second thing is you had to have the person be interested in the area that we wanted. And criteria number three, we needed people who would communicate with others, who would not seal themselves off on their own, who would be good communicators, good collaborators. So this is the way it got started and then we gradually added people.

There was a conscious and explicit decision to build a culture of excellence, founded upon great communicative skills and the ability to work well with others. Kroemer's three criteria guided all hiring decisions. Early on, this meant making hard decisions. Not everyone who applied for the available positions fit the criteria– even though some of them were excellent researchers in their own right. There is no doubt that the strong personalities of Kroemer and Mehrabian and the interpersonal skills of Evans and Merz played an important role in maintaining the ongoing commitment to excellence and collegiality that began to characterize engineering at UCSB. Particularly, for example, in Herbert Kroemer's ability to think dispassionately about any new hire and his willingness to say "no" to excellent prospects if they did not fit with the long-term vision of the school. Much of this orientation can be traced to Kroemer's early experiences of industrial research at RCA Labs and his experiences at Varian Associates. This industrial orientation appears to permeate the UCSB style of engineering. Not only did the school of engineering hire a disproportionate number of their early faculty from industry, but even in the choice of leaders the school has skewed toward individuals with industrial backgrounds. Of the three deans since Mehrabian (who came from a Department of Commerce physical sciences laboratory), two have had industrial laboratory experience (Narayanamurti and the current dean, Rod Alferness) and Mehrabian himself, after leaving UCSB, became president of Carnegie Mellon University and then ended up managing a large industrial concern (Teledyne Technologies).

The other very important decision taken early on that would shape the culture of the engineering college and the tenor of faculty interactions was the decision to ensure that virtually every faculty member in materials also held an appointment in another department. Usually this was assigned with primary affiliation in materials (75 percent) and a secondary affiliation in the other department (25 percent). This ensured a number of very important and useful outcomes. First, it meant that the faculty affiliated with materials were not only limited to graduate student interactions but, through a link to their secondary departments, they could also interact with the undergraduate program. Second, the multiple appointments (and they came with voting rights) meant that the faculty had solidarity with their secondary departments. This practice went both ways. Faculty with primary affiliation in other departments were also granted secondary appointments in materials—for example, Kroemer's primary appointment was in electrical engineering, but he also held a joint appointment in materials by dint of his research area. The joint appointments ensured that materials acted as linking department, diffusing an experimental culture that quickly led to deep, engaged, and productive collaborations across the college. Indeed, for a college of its size (with only five departments) compared to larger members of the American Association of Universities, with typically ten departments in engineering, UCSB has enjoyed an amazing number of collaborative research projects.

The collaborative attitudes fostered by leading faculty such as Evans and Kroemer influenced the culture and intellectual environment at UCSB. A particularly noteworthy example was the recruitment of Art Gossard from Bell Labs to manage the nascent push into Molecular

Beam Epitaxy (MBE). MBE is a computer-controlled method for growing compound semiconductor devices of extremely high purity. It allows for very high levels of control in the process of assembling devices, and was the primary means of growing compound semiconductor materials. The MBE Lab strengthened the school in the area of crystal growth and the new MBE facilities were available to all interested researchers. This had the advantage of further encouraging new collaborations especially among graduate students who interacted with each other in the shared laboratories. The critical mass attained in compound semiconductors led to multiple research collaborations with the semiconductor industry in California and also to the establishment in 1989 of a large NSF-funded Science and Technology Center—the center for Quantized Electronic Structures (QUEST) with Merz as its first director.

The strategic thinking evidenced in the new materials department and the research excellence of the associated hires had a halo effect on the other engineering departments. The influx of new faculty led to crucial changes to science and engineering at UCSB and brought in external resources, excellent graduate students, and new research centers. As previously mentioned, Mehrabian's successors had strong backgrounds in materials-related industrial research, with Narayanamurti (1992–1998) in electronic materials and applied physics and Tirell (1999–2009) in soft condensed matter and chemical engineering. In 1993, a year after Narayanamurti became dean, UCSB became the home of the National Science Foundations' tenth Materials Research Science and Engineering Center (MRSEC), which was housed in a new centrally-located building known as the Materials Research Laboratory (MRL), with strong materials characterization facilities. This in turn led to new collaborations with industry and UCSB today is the home of many government and industrially-funded multidisciplinary centers. Some of these centers were developed during Tirell's time as dean. Of particular note was the opportunity that arose to hire Shuji Nakamura from the Nichia Corporation in Tokushima Japan. Though Nakamura was being actively recruited by other universities, UCSB convinced Nakamura to turn down all his other offers and join the faculty. In 2014 Nakamura won the Nobel Prize in physics for his work on the blue LED. The important question about the UCSB case is what were the factors behind its meteoric rise?

UCSB as a Collaborative Research Institution

One of the most striking elements of the university is the extent to which it was able to escape the departmental silo traps and build truly collaborative research teams and centers—very much like the best industrial laboratories! One of the clearest indications of this collaborative environment was the structure of the materials department and its practice of joint appointments. The strong condensed matter physics effort in the physics department was also an extremely important element in bridging the College of Engineering with the College of Letters and Science. Alan Heeger, a professor of physics, who won the Nobel Prize for his work on organic electronics, was one of the early joint faculty members with the materials department. At a later date, James Langer, a distinguished theoretical physicist and former director of the Institute of Theoretical Physics held a joint appointment in Materials and actually co-taught courses with engineering faculty.

As Matt Tirell, former dean of engineering described it:

we had two opportunities while I was there to populate new buildings. First, a building that had gotten under way but was not built yet by the time [the former dean began the project], but the money was in place and construction started just as I came, for our new engineering and science building, so it was up to me to see who went in it. Later on, with support from the state, we built the California Nano Systems Institute building and the process was repeated in a way. I decided who went into those buildings, and where they went. And, we really did not respect departmental boundaries very much. We really thought about, you know, where would the best synergies be, and asked people, you know, "who would you like to be close to you?"

It's a high-dimensional localization problem and we did the best we could, and we ended up with, you know, there are basically now four big Engineering buildings at UC Santa Barbara, and they all have mixtures of people from different departments in them.

The college's interdisciplinary culture informed the choice of faculty offices by the dean, and these diffuse facilities and distributed resources contributed to even greater collaboration and interdisciplinary work.

Building Shared Facilities and Resources

In the school's early days, the work on molecular beam epitaxy was instrumental in the growing of the new materials and devices used in its research. To facilitate this, the MBE laboratory was from the onset organized as a shared resource. The engineering school at UCSB has devoted a significant amount of resources into building up its facilities to world class standards, and has a tradition of investing heavily in them. One of the deans we interviewed admitted that the school would rather spend grant resources on facilities and equipment than on increasing the size of the faculty, explaining:

> In the major center grants that we received at Santa Barbara or in building planning and setup design, we placed a lot of emphasis on facilities. We had this Materials, Research, Science, and Engineering Centers, MRSECs. It has been going on at Santa Barbara for a while, and we always assigned a disproportionate amount of the budget into equipment and less on the people. Not less—I do not mean that– there is really more spent on the people than equipment, but we could have supported a lot more people, but took the decision to support a big chunk of equipment.
>
> And the philosophy there was, first, you need equipment to stay at the state of the art. Second, if you have certain kinds of good equipment, you know, like electron microscopes and fabrication facilities, you can do a lot of research without even having a research grant. And third, you know, if the funding agency decides to take the money away from you, well, you have some challenges if you have a lot of people to pay, but if you bought a lot of equipment, you still have the equipment. Fourth, facilities

bring people together, you know. Everybody cares about the quality of the electron microscopes or the X-ray scattering or the MBE facilities or whatever. So, I think, investment in facilities is another aspect of space and financial management that catalyzed some good things at Santa Barbara.

This investment in facilities and resources also extended to ensuring that the facilities and resources were adequately managed. For example, upon Art Gossard's assumption of responsibility for the then still nascent MBE laboratory at UCSB, the school was also convinced to hire his research collaborator, namely his Bell Labs' technician. This required flexibility from the school's leadership to hire John English, Gossard's choice to manage the MBE facilities. The dean had to agree to a long-term commitment to the position of "engineer technician" at a rate commensurate to English's seniority to get him to leave Bell Labs and join Gossard at UCSB. English took over the management of the new and growing MBE facilities and did an amazing job of promoting a culture of collaboration and excellence.

Robust Hiring and Promotion Processes

The success of the engineering college at UCSB's was due in no small part to the quality of the faculty and staff that the college hired. As previously discussed, the initial appointments, beginning with Robert Mehrabian's tenure as dean, were senior hires, many of whom had distinguished themselves at industrial research laboratories. However, there was also a lot of junior level hires. In each hiring opportunity, the selection process at UCSB was marked by full engagement of the faculty and the leadership. Each of the three deans—Mehrabian, Narayanamurti, and Tirell, were uncommonly engaged in the hiring of new faculty and staff. As Matt Tirell described in his interview,

So I saw from first hand some of the hiring in the very early days of development of this college. So I'd say, all of the deans took an active role in hiring. And you know, that is not true everywhere . . . and I think that made a difference. I think in the same spirit that we were talking about in connection with other aspects, just general faculty relations, having a dean who was involved and who could talk about his own research as well as what the startup package would be and stuff like that, it made a good impression on people.

In addition to hiring, a particular quirk of the University of California promotion system worked to further strengthen the interactions and inter-relationships between faculty and also with the dean's office. The University of California uses a step system for faculty promotion within the traditional ranks of assistant, associate and full professor. This means that every few years almost every faculty member needs to be evaluated and at the full professor level they underwent very thorough and robust evaluations. This continuous evaluation also required the dean to write memos justifying his or her decision concerning the faculty member's promotion. As it required involvement by the rest of the faculty, the step review process meant that the dean had a very clear picture of the college's research landscape and other members of faculty were kept informed of the work of their peers.

Strategic Long-Term Thinking

As has been described in some detail earlier in this chapter, the culture of the UCSB engineering college was marked by clear strategic thinking and a deep consideration of future possibilities. From the early days, Herbert Kroemer's insight about taking a different path and ignoring silicon for compound semiconductors shaped the college's thinking about how it would approach the question of research focus. Realizing that unlike the much larger and established programs it could not strive for excellence at everything, resources were devoted instead to a few critical areas where the school *could* achieve excellence. For example, given its size, the engineering college at UCSB only has five departments (chemical engineering, computer science, electrical and computer engineering, materials, and mechanical engineering). There are no departments of civil engineering, nuclear engineering (the last vestiges of the nuclear engineering program were dismantled in 1995), or aeronautical engineering. One possible explanation for why this mind-set proved so successful was the relative youth of the school at the time that it began to grow. It didn't have decades upon decades of institutional culture to upturn and change. While this was undoubtedly true and played a role in the wholehearted embrace of strategic thinking, another important element that allowed UCSB's engineering school to act strategically was the strength of its early leadership. All three deans since Mehrabian have been able to provide leadership without riling up the faculty. They each found a balance between issuing top-down directions to faculty and responding to bottom-up initiatives from the faculty. For example, it was Matt Tirell's enthusiastic response to a faculty initiative that led to Shuji Nakamura's hiring. This enabled them to lead the faculty and the rest of the school toward a common goal. Also, the presence of inspirational faculty members (such as Herb Kroemer and Tony Evans) who actively promoted the strategic vision as a key ingredient of a culture of excellence and collaboration cannot be underestimated. These various elements contributed to a relatively lean engineering school that was able to rise to a place of international prominence in a very short span of time.

External Engagement

The engineering college at UCSB established and sustained engagements with a number of important external communities that were critical to its success. Much in the same way that one cannot fully appreciate or understand the work at Bell Labs without considering its position within AT&T, its relationships with the broader academic community, and its close ties to the needs of Western Electric. In the same vein, understanding the culture of the engineering college at UCSB requires an appreciation of the various kinds of external engagements that helped drive its work of research and teaching.

Chief among them were the many collaborations across the UCSB campus and with the surrounding community. UCSB is a relatively small campus and is one of the few members of the American Association of Universities that does not have professional graduate schools in business, architecture, law, public health, or medicine. Consequently, the engineering college at UCSB does not compete with any professional school for resources or prominence and, partly as a result, has forged strong relationships with the sciences. As mentioned previously, faculty in physics have held joint appointments in engineering. This is also true for other departments and programs across the university. Also, the absence of a business school has meant that the engineering college has developed its own culture of entrepreneurship, reaching out to the local business community. Indeed, the engineering college's impact in the broader business community is so influential that Dean Narayanamurti received an innovator award in 1997 from the local chamber of commerce for his work as dean of engineering and for encouraging an entrepreneurial culture in Santa Barbara. Entrepreneurial activities at UCSB engineering have coalesced into the Technology Management Program and UCSB is currently ranked very high in startup activity. For example, UCSB beat Harvard and the University of Pennsylvania (places with preeminent business schools) in Forbes 2014 rating of entrepreneurial schools.[5] UCSB also boasts enviable access to venture capital funding. Dean Rod Alferness explained to us that:

> I should mention the entrepreneurial spirit here and the fact that our faculty fundamentally want to have impact. One of the ways of doing that is through startups. Our statistics indicate a very high number of startups per hundred million of research funding. I believe we lead our UC sister universities on that metric and we rank very highly overall.... Our students got very involved during the telecommunications bubble when startups were all the rage and in response to the increasing demand from the students we started the Technology Management Program. That's really about helping our engineering students understand some basic principles about business and finance, et cetera. But fundamentally help them to understand that quite frankly, in order for technology to be used it's got to be commercialized. In order for it to be commercialized, there has to be a value proposition of why someone would want to put in an investment in to realize the technology.

In addition to its work on technology entrepreneurship, UCSB also has a number of very important industry-university collaborations that have helped drive the research programs in engineering. Three very interesting examples are: the Solid State Lighting and Energy Electronics Center (SSLEEC), a center that partners with multiple industry partners; the Mitsubishi Chemical Center for Advanced Materials, (MC-CAM), an interdisciplinary center that brings together UCSB engineering faculty strength in materials and other physical sciences faculty with engineers and scientists from Mitsubishi chemical; and, finally, the Dow Materials Institute. All of these are unique collaborations. The university has had to be very flexible in structuring these relationships to protect faculty and graduate students from conflicts of interest while simultaneously encouraging cross-fertilization and innovation.

Janelia Research Campus: An Experiment in Doing Research

Janelia Research Campus is the research arm of the Howard Hughes Medical Institute (HHMI). HHMI is one of the most respected and prestigious medical research funding organizations in the world. Headquartered in Chevy Chase, MD, HHMI supports individual researchers at universities with very generous funding packages. In the community of medical research funding, there is perhaps no more reputable brand and respected category as an HHMI Investigator. The endowment of HHMI grew substantially in the early 2000s and under the leadership of Tom Cech the institute decided that instead of increasing the number of individually funded investigators, capped at about 350, it would experiment with a different model. The decision was made to create a novel institution that would bring researchers together in a central research campus with the aim of having a large impact on biomedical science.

In 2000 Gerald Rubin (a very successful HHMI investigator) was recruited from the University of California at Berkeley to become the new vice president of biomedical research with a broad mandate to build a new institution. Rubin's history included time spent as a PhD student in the United Kingdom at the University of Cambridge working at the renowned Medical Research Council's Laboratory of Molecular Biology (LMB). The LMB at Cambridge had established itself in the 1960s and 1970s as an influential center of medical research. Funded primarily by the Medical Research Council—the leading public-funded biomedical research organization in the United Kingdom—the LMB's budget is set every five years and is flexible within that period, allowing similar sorts of intellectual freedom enjoyed by researchers at Bell Labs. The sustained funding also allowed the LMB to tackle difficult problems in biology. Finally, the proximity and close connections to Cambridge University contributed to the LMB's excellent research culture.

LMB was the workplace of numerous renowned researchers such as John Kendrew and Max Perutz (who discovered the structure of proteins), and Fred Sanger (who won two Nobel Prizes in chemistry, the first for his work on sequencing insulin, and the other for sequencing DNA). The LMB was also the workplace of the world famous Watson and Crick, and various other Nobel Prizes laureates, and had a global reputation for groundbreaking research in the biological sciences. Rubin's experience at the LMB was an important motivator for the new research campus that he would ultimately build.

After taking the time to study the challenges of institution-building and reviewing other very successful medical research institutes, in particular MIT's Whitehead Institute and the Salk Institute in La Jolla, a decision was taken by the HHMI leadership to do something different. Using the LMB and Bell Labs as templates, HHMI, with Gerald Rubin providing leadership, decided to invest substantial resources into building a new kind of institution. The thinking was that the institute would, as had occurred at Bell Labs and the LMB, insulate researchers from other responsibilities such as teaching and grant writing, allowing them to devote all their energies to the singular pursuit of research. In a way, the new institute would be a grand sociological experiment, an intervention to change the way *research* in the biomedical sciences and engineering was undertaken. Rather than an institute setup to focus on a particular problem or area of research, HHMI decided that the selection of the biomedical problem that the Janelia Farm Research Campus (later renamed Janelia Research Campus) would focus on would come much later. First, the fine details concerning the institute's design and philosophy were worked out.

By design, Janelia was unique in that it brought together the biological sciences with the physical sciences, computer sciences, and engineering in a new kind of research institution that would benefit from stability in funding and be structured to address newly emerging areas in biomedicine. The new institute would bring together, for the first time in HHMI history, biological scientists, engineers, and physical scientists all under the same roof and all focused on resolving a big research problem. This point alone makes it a fascinating and apt case study for our investigation of institutional design. Janelia was in many ways, a reaction to the perceived limitations of the traditional academic model of research and an embrace of an industrial model. Gerald Rubin identified several aspects of the academic model that the Janelia Research Campus was expressly designed to counter. For example, traditional academic research laboratories work on the model of the professor as laboratory administrator. The professor provides high levels of guidance and mentorship, but the actual research is carried out by graduate students and postdocs. Most assistant professors in their very first jobs could quickly go from being active participants in research to being expected to take up a mentoring and administrative role, hiring graduate students and postdocs to do the actual work. This often happened at the most active and creative point of their intellectual trajectory. At both Bell Labs and the LMB, active researchers were just that—active in research. They didn't manage other people who did the work, they did it themselves. By adopting small laboratory sizes and not having the pressures of service, teaching, or fund raising, Janelia was expressly designed to allow researchers to take a hands-on approach to their work in an intensely multidisciplinary environment which necessitates a collaborative culture.

Also, as HHMI was devoting two billion dollars over a ten-year timeframe, the research campus would have the resources to tackle complex problems that could not be tackled by any single investigator. Having an institution with financially empowered leadership also meant that the research campus could respond quickly to advances or to challenges, bringing resources to bear in an efficient manner. In its program development report, Janelia Research Campus also identified collaboration as a challenge within the traditional academic model.[6] The report argued that the academic incentive and reward systems (tenure and promotion) discourage collaborative work as junior colleagues obsessively sought to catalogue their individual contributions in order to strengthen their case for unique work output. High-risk high-reward collaborations were consequently inhibited. In essence, Janelia Research Campus was designed from the ground up to address some of the limitations of existing models of academic biomedical research funding, and was an attempt to adopt best practices from the LMB and Bell Labs. The goal was to take a long-term view to the funding of transdisciplinary research, creating a deeper integration of biological, engineering, and physical sciences disciplines. Janelia Research Campus opened its doors in 2006, and the past eight years provide us with a record with which we can begin to assess its promise.

It is one thing to build an experimental space for investigating new and unique research organizational forms. It is another thing to successfully build a new model for research. To undertake our case study of Janelia Research Campus, we visited the research campus and conducted interviews, from members of the executive team to laboratory heads to postdocs and support staff members. Our goal was to gain an understanding of how the research campus functioned and actualized some of the principles that were so well articulated in its founding documents. We sought to understand how Janelia was translating the Bell Labs experience and the LMB experience in an institution designed to be radically different but sitting within one of the most successful funders of the traditional model—HHMI.

Building a Collaborative Research Institution

One of the most striking things that will confront any visitor to the Janelia Research Campus is the institute's physical architecture. Designed by the internationally renowned architect Rafael Vinoly, the institute is a high-tech paradise of glass, wood and concrete that communicates an aura of wealth, sophistication, and attention to detail. Tucked away in a quiet corner of Ashburn VA, in Loudon County—the wealthiest county in the United States—the Janelia Research Campus is located a drivable distance from the

HHMI headquarters and from the National Institutes of Health. The building is made up of three floors above ground with a single corridor running through each floor. The liberal use of glass in its design lends a sense of openness and transparency and allows the residents to see at a glance which laboratories are occupied and who is present.

Taking cues from Bell Labs (Murray Hill was famously one building) and the new LMB building in Cambridge (which also has three main floors), the single building is arranged in two wings—east and west with a central section of equal size in between. There is a central café on the ground floor that offers excellent fare and hours that encourage everyone to have lunch at roughly the same time. The institute also has extensive underground laboratory space, and the grounds contain substantial residential facilities giving Janelia more of the feel of an impressively endowed academic campus than an industrial research laboratory. Given the high cost of living in Loudon County and the isolated nature of the city, the residential and dining facilities at the institute combine to encourage Janelia researchers to remain continuously on site.

The physical building is optimized to maximize serendipitous interactions in the main corridor and cafeteria. The clear glass walls ensure that other researchers are never fully out of sight—facilitating drop-by conversations. Taken together, these features ensure that Janelia Research Campus is a place where all that could be done has been done to ensure that physical design reflects the integrative and collaborative culture that the institute was designed to embody. Few places in the world have so carefully thought through and integrated their espoused values into their physical environment.

Of course, all the inspired design in the world does not make an institutional culture, and without clear policies and incentive structures, the physical plant alone would not lead to collaboration. Just as important (or more) in promoting a collaborative environment is the tone set by the leadership. In this, it is clear that Janelia has made enormous strides. In response to the question "How would you describe the research culture at Janelia?" all our interviewees would describe it as collaborative. As one of our interviewees who had previously worked at Bell Labs recounted:

It sways very much toward collaboration—very much—and one of the metrics by which you are judged here is how collaborative you are. Internal competition is really almost I would say nonexistent in a direct sense. There might be some competition in terms of only the competition which—and which is the same competition I felt at Bell—is you want to show you are at least as good as your peers. . . .

By making collaboration an integral aspect of the evaluation process, the Janelia leadership has succeeded in institutionalizing a collaborative culture. One interesting observation from the quote above that was echoed by others is the idea that competition is not valued as highly as collaboration, as was the case at Bell Labs. Sure, there was competition between projects—for example, Janelia has multiple projects devoted to building next generation microscopes, each using different approaches. In addition, the institute has at least two major research programs organized around different model organisms. The east wing of the building is nominally devoted to drosophila (fruit fly) research and the west wing is also nominally devoted to mice research, with the center being the "demilitarized zone" to echo a term that came up frequently, there is the sense that in general, competition should be deployed in the service of sustaining the collaborative environment. As one of our interviewee's put it: "If there is competition, I think there is a strong effort here to ultimately make it the selfless agnostic kind of competition." It is also clear that the emphasis on collaboration has had the effect of virtually ending the formation of disciplinary based silos. The institutional culture at Janelia is organized around projects and what people do, not from which department they received their terminal degrees. As another interviewee said: "Here people are interested in what you do and work on, not in what your degree is in."

Building an Elite Institution

Anyone spending any appreciable length of time at Janelia Research Campus is left with a clear impression that Janelia, from its impressive building to the people who work there, is an elite institution. All high-performing and innovative research institutions will cultivate, promote, and protect a sense of being elite. This does not imply an aggrandized sense of self-worth, but rather, a sense that brilliant people work here, which translated for individuals as, therefore I must push myself to belong. Though it can cause "impostor syndrome" or the sense that one does not belong and will be found out as a fake by one's colleagues in newcomers, it is this very sensibility that draws the best out of everyone and in many cases encourages a sense of humility. The best institutions cultivate this impression of being elite and use it to highlight the importance of the work being done and the need for a communal effort. Bell Labs for example, was renowned for its ability to quickly imprint on newcomers and visitors alike, the sense that only excellence was acceptable, and everyone who was privileged to work there was a superstar. Such a culture can endow a healthy confidence that is essential to researchers' ability to take risks and ask interesting questions.

Janelia has done a great job in building a sense of purpose and competency and it is reflected at all levels from the director to the postdocs. This is especially remarkable given the relatively short time since the institute's formation. As one of our interviewees put it:

> So I guess I would say people here are, like at Bell, are generally way above your normal—you go to any academic department and then give a talk and you go talk to people during the day, the average here is way higher than the average in any Harvard department I've been to or anything like that.

A large part of this sense is clearly derived from Janelia's association with HHMI, and the intellectual prestige affiliated with the HHMI brand. As another interviewee put it:

> But in terms of the institution's reputation there is no better brand in life science than HHMI. It is as good as it gets because, as I said, the principal people know that the 300 best biologists in the United States are HHMI investigators. We collect Nobel Prizes like there is no tomorrow. And so, if I call somebody up and say, "This is (James) from HHMI," or send an email saying

this is (James) from HHMI, I've got that stamp on my forehead
. . . instantaneous credibility.

it is also clear that the very generous resources available to Janelia
contribute to the feeling that the place is something special. The resources also create a sense of responsibility and a pressure to perform
at an elite level in order to justify the generous allocations. As one of
our interviewee's put it:

> I think most people do feel that because of the freedom and level
> of resources that we have, there's sort of a similar belief that we
> can't just do research that's competitive with even, say, the top
> quartile of what's out there. But we have to be exceptional in
> what we do or else it's hard to justify what we're doing.

These resources are not only budgetary, but may also be found in the
support systems that are all oriented to facilitate research. This finds
tangible expression in the desire of the Janelia administration to do
everything in its power to remove any and all obstacles to research.

Optimizing Administrative Support

All the Janelia support staff members (including very highly placed
administrators) were clear that the most important aspect of their
job was to facilitate research, and they were all eager to communicate
their willingness to solve any problem in support of the research mission. A newly appointed postdoc from Harvard had this to say about
his experience at Janelia:

> The thing is that they set up this place so you really feel like
> everyone here, not just the postdocs or [PI's], everyone here is
> here to make the research happen. So, they have facilities to
> make cell culture, to take care of animals. Even the guy who
> cleans the room, or set up security procedures are here to make
> your job easier, to make the research job easier. For example,
> when we set up a new room for microscopes with lasers, the
> security folks came to our assistance to check out things for us

and asked if we needed laser safety goggles or whatever equipment, and they purchased them for us. So you feel like they are
not trying to make their jobs easier, but they're here to make
the research easy.

In a similar vein, a lab head testified to the orientation of the support
structure at Janelia:

> Productivity also comes about by administration here. [Gerry
> Rubin] was very frustrated when he was at Berkeley by all the
> red tape and he felt that there were people all through the ad-
> ministration who took an active role in actually suppressing the
> ability to get stuff done. And so he swore when he was going
> to design this place that people over on the first floor at that
> end [administrative end] understand who is in charge and who
> is in charge are the researchers and they are there to serve the
> researchers. I have had two postdocs leave in the last year and
> come back and just cry, cry, cry about how even though they
> have all the start-up money, [. . .] how difficult it is to spend it.
> And here it's just so easy. They just knock down all the barriers
> to make that happen. It just feels to me like everything I see in
> academia elsewhere is in total slow motion to the way we work.
> It takes some years to do stuff that takes us weeks.

It was clear during our time there that the support systems at Janelia
Research Campus contribute significantly to the very special environment that the institute enjoys. Much of this seems to be as a
result of the stated and much repeated policies of the upper level
administration. The understanding that all parts of the institute are
laser-focused on facilitating the research mission was communicated
clearly to us by everyone we interacted with who was in a support
role. As another interviewee, a lab head put it,

> I mean, we are way, way ahead and it is because of the structure
> that Gerry helped create here. I mean, it is having not to worry
> about money. It is having an ability to work in the lab yourself.
> It is the ability to keep a group small. And it is the ability to

have all of the rote stuff done by others so that your people can concentrate on the things that they are truly expert in, right?

Shared Resources

Janelia, like a number of other very successful research laboratories of the past (Bell Labs and the LMB, for example) has established a culture of shared resources. While there are many project resources that are unique to a particular project, there are a lot of shared resources that are available to all the project teams. For instance, all the fly management and mouse management facilities are centralized in thoroughly modern and adequately staffed laboratories. This allows the individual teams to remain small and off-load the mundane responsibilities of managing modern laboratory animal and insect models to a specialized staff.

> The shared resources definitely promote interactivity . . . There is a lot of rote stuff in biology about keeping cells happy or flies happy and so there is all those shared resources to do that. So they are kind of the after-burner that allows you, in the small group, to get productivity well in excess of what that group by itself is capable of doing. So that is a huge boon.

Apart from increasing the productivity of individual groups, the shared resources also act as meeting places for various group members to interact and spark cross-pollination of ideas across groups. The engineers and scientists that act as staff persons for managing and providing these resources act as bridges between various projects, circulating knowledge and expertise:

> So the fly, there are thousands of genetic variants of flies, and they get serviced and housed robotically and fed and cleaned and bathed, and I don't know, what else, you know, maintained in various ways. There is also a vivarium for the mice, and there's a certain level of scientific services there for people to not have to deal with the more routine aspects of it.

These shared resources are not just biological tools. Janelia has devoted a significant portion of its budget to tool building—both biological and physical tools. While the east and west wings are devoted to fly and mouse research, respectively, the center wing is given over to imaging research and to the applied physics group. The imaging research teams have their own lab heads and operate in a similar fashion to the other research teams at the institute with their own postdocs and other scientific members of their labs. However, they are also a centralized resource for all the other scientific groups as they work on inventing and building cutting edge devices to capture images of biological samples. Janelia's 2014 Nobel Prize came from this group. The engineers and physical scientists in the center wing also help in circulating knowledge and provide institutional memory of the various strategies and processes they invented to meet the unique challenges of the researchers at Janelia Research Campus.

Focus on Competitive Advantage

So, I would say that thing that makes Janelia, as far as I can tell, unique in that sort of twentieth or twentieth century, is it is the only research institution I know that was set up with a goal of doing science in a different way rather than to solve a particular problem. . . . In fact, I describe Janelia as a biotech company whose product is new basic knowledge with infinitely patient investors.

—Gerald Rubin, Executive Director,
Janelia Research Campus

It seems the most productive research institutions have an ability to say no to projects that don't line up with their competitive advantage. Thinking strategically about what the institution is uniquely placed to contribute is crucially important in determining the direction the research should take. At the onset Janelia went through a process of determining what its focus would be. As the quote above shows, there was an early acknowledgment that the institute would above all, do science differently. How that played out in the current incarnation of the institute (and Janelia is setup so that it is capable of changing research directions completely if a decision is made to do so) is worth examining in detail.

One of the institute's unique qualities is the laser-like focus on research that the overly generous resources allow. There are little to no distractions. There is no grant writing, there are no service requirements, and there are no teaching responsibilities. The sole purpose is research. This allows Janelia to recruit some very competitive individuals. The lure of a pure research agenda has resulted in several individuals giving up tenure at some of the very best institutions in the country to accept positions at Janelia; people have also left positions in industrial research. The institute also asks the same questions of every project: "who else is doing this?" and "would it get done if there was no Janelia Farm?" If the answers are no one else is doing it and it would take the unique resources of Janelia to accomplish it, then the project usually receives a green light. The goal is to have the highest impact possible. At the onset, Janelia's leadership engaged in conversations with the NIH director, and over multiple planning sessions arrived at the realization that imaging was a core area of need in the biomedical community. No one else was doing it in a concerted way and Janelia had the resources and the heft to get involved. The 2014 Nobel Prize in chemistry was given to Eric Betzig, an applied engineering physicist working at Janelia on imaging, who graduated from Cornell University's College of Engineering. While the Nobel was given for work carried out before Janelia employed him, Betzig was unemployed before Janelia hired him and gave him the resources and environment to continue his work.

The other area that was identified early on as a focus was the need to understand how the fruit fly (drosophila) brain works as a step in understanding the human brain. This tied in well with the focus on imaging and these are the two major current areas of focus for Janelia.

The laser-like focus on research is also carried over into personnel policies. As one lab director told us:

> Far and away the most important one is the sort of mental space and time to think about things. I am encouraged not to travel, for example. Its policy is that we have to be physically present at Janelia 75 percent of the time, and it is not like they are watching us with punch clocks or anything, you know, there is just an expectation. And that is useful because when I get invited to

do things, I use it as an excuse. I say, "You know, I would love to come and do your thing, but [Gerry Rubin] won't let me." That's great!

This ability to pick a niche that is unoccupied and pour resources into it appears to be a characteristic of high performing research institutions. Bell Labs, for example, owes it genesis to the need by AT&T to solve long distance amplification of communications signals (necessitating the vacuum tube) and then later on the need for a solid-state replacement for the vacuum tube (leading to the invention of the transistor). Focus is essential in directing the resources of research institutions.

One area in which Janelia was able to leverage its unique resources in a competitive way, and accomplish something that only Janelia could have done, is its project on biosensors. Named GENIE, the biosensors project is one of the major success stories of Janelia Research Campus. The GENIE project has led to new classes of biosensors, judged by the metric of how widespread their work has become, it has been a smashing success: this project has achieved close to 95 percent market share. The reason is simple. Broad impact came because of the major improvements in protein biosensors that the GENIE project accomplished. Janelia Research Campus could do it better than anyone else, because it required detailed, repetitive, tedious work and was frighteningly expensive (roughly $2 million a year for five years). It was clearly a project that wasn't being done, and Janelia was able to apply the necessary resources and scale up quickly. In other words, it was a perfect fit for the institution with a clear payoff in research impact and widespread adoption.[7]

Building Connections to Relevant External Communities

Research institutions cannot operate in a vacuum. The fruits of the research need to be shared with others. Interesting problems and insights come from industry and from helping solve other people's problems. Research is not a solitary activity, but rather one that is embedded within relevant communities of practice. Janelia realized that, to be impactful, it has to circulate people and ideas. Given its relatively isolated situation this has been addressed in different ways.

To increase and coordinate Janelia's research activities with relevant academic communities, a visitor's program has been created that makes funds available to bring in collaborators. As one lab head described it to us, the visitor program is a large part of the reason why he is able to keep working at Janelia:

> But, on the other hand, the visiting program here has been a secret weapon for me . . . HHMI has the network that Bell had in condensed matter physics back in the day. Biomedical science, if you are in HHMI, you are in a fraternity of the best of the best, right? And so between that and the visitors program, it's been very easy for us to get other researchers to come here. Instead of having to go out, we can have people come here. We have had thirty-five different groups in the last year come to work in our lab for at least a week at a time each. So that's kind of saving my bacon or else I'd probably be looking outside already for something else.

In addition to the visitor's program there are major biomedical conferences held at Janelia that attract the top researchers from all over the world. These and other events held at the institute promote its work and ensure that it is at the center of the biomedical research community. As a top administrator described,

> "We have probably a thousand people that come in and out. We have the conference programs. We have about fifteen conferences every year. And they're usually organized by a mixture of people from the inside and the outside. We have maybe a whole bunch, probably dozens and dozens of visitor projects where we have a pot of money, and that's one of the things that I do, where we encourage collaborations between somebody on the outside and, ideally, one or more people here.

The interconnections are not only limited to other research institutions, but there are also productive ties to industry. Some of the inventions in imaging are on their way to being commercialized.

Common Elements in Research Excellence

The two examples above could not be more different. Janelia is a philanthropic-funded research institute that does all that it can to facilitate an insulated, well-supported environment for the very top researchers in the biological sciences. UCSB is a university with an exceptional college of engineering with a clear focus on materials that has leveraged the resources it has into worldwide prominence and impact. Both institutions were honored with Nobel Prizes in 2014. Obviously, these places are doing something right.

On close inspection, it is clear that they share many common elements which are critical for the flourishing of the discovery-invention cycle as enumerated in Chapters 5 and 6. They have both evolved cultures of excellence based on similar precepts. Perhaps most important, they are both driven by unique and focused strategies that embrace a vision of how research distinction can be achieved in selected areas through building a critical mass of engaged and cooperative scholars. Leadership at these institutions is closely tied to the work of research; they both promote and embrace collegiality; both institutions have found unique ways to bring people together in interesting ways; and both have institutional safeguards to protect against the creation of silos of people and ideas. Both have invested heavily in infrastructure and pay very close attention to hiring and promotion seeking excellent hires who would fit with their institutional cultures and values.

8

THE NEED FOR A RADICAL
REFORMULATION OF S&T POLICY

WE HAVE OFFERED the discovery-invention cycle as an alternative to the problematic construction of "basic" vs. "applied." We have also examined the necessary institutional culture and arrangements, exemplified by the great industrial laboratories that promote a discovery-invention environment over the silos and conflict that come about partially as a result of the inherent divisions in the "basic" vs. "applied" framework.

The last chapter demonstrated that it is possible to build contemporary research institutions that implement the structure and culture necessary to promote a virtuous cycle of discovery and invention. Two exemplary institutions, UCSB and Janelia Research Campus, have given homes to and nurtured some unusual Nobel Prize winners whose work was awarded the prize in the last decade. The need to replicate such institutional arrangements on a national scale and in our public mission-oriented agencies like the Department of Energy and the National Institutes of Health is more pressing than ever before.

The call for change directed at national science and technology policy is not entirely a new one. Several blue ribbon panels have advocated for change in the United States S&T policy as have the various national academies. This need for change and a desire for science, engineering, and technology policy at the highest levels to move in a different direction has been expressed in various ways. In

130

this chapter we examine a number of these prescriptions. Given the pressing challenges that the planet has to confront, and which must be resolved with the aid of science, engineering and technology, it is imperative that we get things right at the very highest levels of policy making in the halls of Congress, at the executive offices of the president, and in our national funding institutions. The truth is that even if we adopt a shift in language and implement comprehensive cultural and structural change at various research institutions, if there is no change in US science, engineering, and technology policy at the very highest levels, all other efforts will be stymied.

We begin our analysis with the 2006 report by the National Research Council (NRC). Chaired by Norman Augustine, an eminently distinguished aeronautical engineer, the report ominously titled *Rising above the Gathering Storm: Energizing and Employing America for a Brighter Economic Future*[1] laid out four recommendations focused on K–12 education, research, higher education, and economic policy. These were further broken into twenty concrete implementation steps.

The report was exemplary, not only for the very short amount of time in which it was completed, but also for the widespread bipartisan reception with which it was received. Many of the report's recommendations have since become law, for example, the "American Competitiveness Initiative" and the "America COMPETES Act." The most recent recommendation that was finally actualized was the Advanced Research Projects Agency-Energy (ARPA-E) which was funded by President Obama almost three years after the publication of the initial report. Though the 2011 update to the report *(Rising above the Gathering Storm, Revisited: Rapidly Approaching Category 5)*[2] was critical of the general outlook of the nation with regard to implementing the previous recommendations, by the standards of change in the federal government, the initial report can be seen as highly influential and led to a number of institutional changes on the various fronts described in the report. There is much to consider in this report. Of particular relevance to the discovery-invention cycle is the clear link between education and research that the report explores.

It has become increasingly clear to us that if the ideas behind the discovery-invention cycle are to become widespread a few important changes are necessary in how we value theory and practice in our educational institutions.

Education, Theory, and Practice

In his book *Fragile Objects*[3] the French Nobel Prize winning physicist Pierre-Gilles de Gennes is highly critical of the French system of educating scientists. The meat of his criticism is clearly discerned in his account of his interactions with graduates of the Polytechnic School of Paris who came to Orsay, France to attend an advanced program in solid-state physics. De Gennes' would set them a practical problem having to do with the setup of a simple cosmic ray detector. This was a problem that their prior training should have adequately prepared them for. However, a large number of the students were unable to consider the problem outside of their customary specific theoretical frameworks. In de Gennes' own words:

> "On their final exam, I would give them a problem of the following type: "Imagine a thin, evaporated metal film, like lead, 1 micron in thickness. A cosmic ray with an energy of 10 MeV (megaelectronvolt) traverses the film, which is held at a temperature of 4 kelvins. A voltmeter is connected between the edges of the film. What are the amplitude and duration of the resistance pulse that can be measured across that film? Is it possible to use this design to build a simple cosmic ray detector?" . . . Two or three general concepts of this kind are all that are needed to predict the amplitude and the duration of the thermal pulse. But the typical Polytechnic graduate I inherited at the time would remain stumped in front of his bare blackboard. One of them finally blurted out (I will never forget his comment!): "But, sir, what Hamiltonian should I diagonalize?" He was trying to hang on to theoretical ideas which had no connection whatsoever with this practical problem. This kind of answer explains, in large part, the weakness of French industrial research." (*Fragile Objects* p. 156)

This inability to translate extensive and comprehensive theoretical understanding into practical consideration is a problem de Gennes ascribes to the work of Aguste Comte. Comte, one of the founding fathers of Sociology and the Philosophy of Science famously promoted a classification of science in his positive philosophy.[4]

Positivism is a philosophy that holds the ability to exactly determine phenomena as the most vital criterion in the furtherance of any field of study and thus mathematics and physics are more positivist than chemistry or biology. While it is possible to read Comte without imputing any value judgments about the relative merit or demerits of the more positivist sciences (like mathematics and physics) over the sciences that have to account for ever increasing complexity (for example sociology) it is clear that the more mathematical sciences such as physics and astronomy have come to be generally seen as more rigorous and exacting than the more descriptive sciences such as biology and chemistry.

For de Gennes' the French educational system's move away from experiential learning connected to actual problems, to more theoretical and mathematical instruction was a clear by-product of the valuation of theory and abstraction over the ability to think in physical terms. A state of affairs he pejoratively described as "positivist prejudice." Though de Gennes' criticisms of positivism as the bane of thinking in physical and practical terms might be a bit extreme, there is truth to the broad observation that in the present era, theory and abstraction are often considered more highly than experimentation and observation. There is a strong allure to attribute greater value to the imagined increased abstraction of "science" over the inadequate colloquial understanding of "engineering" as a practical art.

If there is one key intervention we hope this book will have, it is that it will spur greater conversations about how we sometimes value the abstract over the material in the design of academic curricula, in our science and technology policy, in the design of our research institutions and in the halls of academia. Perhaps it is time to grant equal importance to both abstraction and concreteness. This is especially necessary at the present time where pressing global challenges require transdisciplinary efforts. The National Academy of Engineering's Grand Challenge Scholars program aims to develop a new cadre of engineering graduates whose curricula training clearly connects with specific grand challenges is a highly promising development in this area. The participation of 120 engineering programs as of 2015 bodes well for the future of engineering education in the United States and for Discovery-Invention Cycle thinking. In a similar vein, in the field of chemistry, George Whitesides and John Deutch call for an overhaul in the practice of chemistry. "A focus on the practical does not mean ditching fundamental science. It means using fundamental science for a purpose and practical problems as a stimulant to curiosity. Chemists can still be curious, en route to addressing the big societal challenges of our times."[5]

Rethinking S&T Policy at the Highest Levels

As we are interested in the import of the discovery-invention cycle on research broadly described, it is important to turn now to the arguments in *Rising above the Gathering Storm* that deal with the research environment in the United States. The general recommendation concerning research was clear in its ambit:

> "Sustain and strengthen the nation's traditional commitment to long-term basic research that has the potential to be transformational to maintain the flow of new ideas that fuel the economy, provide security, and enhance the quality of life."

This was followed by six implementation action points for the federal government:

1. Increase the federal investment in long-term basic research by 10 percent each year over the next seven years through reallocation of existing funds or, if necessary, through the investment of new funds . . . . Increasingly, the most significant new scientific and engineering advances are formed to cut across several disciplines.
2. Provide new research grants of $500,000 each annually, payable over five years, to 200 of the nation's most outstanding early-career researchers.
3. Institute a National Coordination Office for Advanced Research Instrumentation and Facilities to manage a fund of $500 million in incremental funds per year over the next five years.

4. Allocate at least 8 percent of the budgets of federal research agencies to discretionary funding that would be managed by technical program managers in the agencies and be focused on catalyzing high-risk, high-payoff research of the type that often suffers in today's increasingly risk averse environment.

5. Create in the Department of Energy an organization like the Defense Advanced Research Projects Agency (DARPA) called the Advanced Research Projects Agency-Energy (ARPA-E).

6. Institute a Presidential Innovation Award to stimulate scientific and engineering advances in the national interest.

(Rising Above the Gathering Storm, pp. 7–9)

These six action items are specific and the report goes into significant detail as to how each should be carried out. Yet it is clear that only a few of these points actually address the structure of research activity in the United States, and the report falls into the "basic vs. applied" trap. The first implementation point highlights the need for long-term "basic research" in both science and engineering, with the key observation that advances that cut across disciplines are especially worth supporting.

The overall thrust of the report with regard to the research environment remains primarily concerned with increased funding for science and engineering "basic research." Little is said here that addresses the kinds of challenges we have called out in this book and only a few of the recommendations listed above were adopted and of those few the adoption was only partial. There was an increase in federal investment, but not at the level suggested. The early-career program exists today, but, again, not funded at the level suggested. Perhaps the greatest achievement has been the creation of ARPA-E, which interestingly enough holds the greatest promise to change research culture by addressing critical structural challenges to creating new energy technologies.

ARPA-E is a significant achievement, however, it has become increasingly clear that ARPA-E may not achieve its full potential as its funding is subject to the various short-term political winds in Washington, DC. There are constant battles[6] over the level of funding for the organization, making long-term planning and research execution difficult. Part of the problem is the framework with which policy makers think about research. The "basic" vs. "applied" framework that policy makers utilize has resulted in significant resistance in Congress to funding "applied" research on the false belief that it is the appropriate responsibility of industry. We have shown that this is simply not the case. Considering research as a dance between discovery and invention would lead to a very different orientation on the part of Congress. Considering research as a holistic endeavor means an appreciation of the interconnectedness and indivisibility of the research taking place in related Office of Science programs, and the research undertaken by ARPA-E. As we have argued, there is a boundary to be drawn, but it should be between research and development. Even then, the truth is that *even* this boundary between research and development cannot be absolute, but rather should be managed to ensure the best possible outcomes where the researchers performing the work need to have the freedom to move across the boundary.

Institutionalizing Multidisciplinary Environments

The second policy prescription that we will review are the two recent American Academy of Arts and Sciences reports dealing with Advancing Research in Science and Engineering *(ARISE)*[7]. The first *ARISE* report was chaired by Thomas R. Cech, a distinguished biologist who once headed the Howard Hughes Medical Institute. That report, unlike *Rising above the Gathering Storm*, did not focus on increasing funding levels but rather on improving US leadership in science and technology by supporting early-career faculty and encouraging high-risk, high-reward transformative research.

> We strongly believe that, regardless of overall federal research funding levels, America must invest in young scientists and transformative research in order to sustain its ability to compete in the new global environment. In this report, we outline a series of recommendations for all key stakeholders, including government, universities, and foundations. (*ARISE*, p. 1)

ARISE broke down these two goals into clear recommendations for universities, private foundations, and federal agencies. The major message was the need to increase funding to early career researchers and to high risk proposals. The authors were critical of the increased grant writing burden placed on early career faculty and the low probability of funding especially in grant review processes that tended to revert to the mean and thus underfund high-risk, high-reward projects.

For our purposes, the primary policy prescription of the 2008 *ARISE* report dealt with the research funding side, however, it did not address the structure of the research environment. What it did do very well was address the challenge of interdisciplinarity, particularly as it applies to the physical, biological, life sciences, and engineering, and raise the need for enhanced funding for young investigators in the NIH's grant-awarding processes and in the grant-making process of private foundations.

The 2013 *ARISE II* report, subtitled *Unleashing America's Research and Innovation Enterprise*[8] was chaired by Keith Yamamoto and Venkatesh Narayanamurti, an author of this book. More important, Narayanamurti is a physical sciences researcher, and Yamamoto, a distinguished molecular biologist. This is the first report that effectively utilized the term "transdisciplinarity"[9]. *ARISE II* defines transdisciplinarity as the ability " . . . to integrate fields beyond the levels of the multidisciplinary, in which multiple disciplines operate simultaneously, or the interdisciplinary, which occupies the space between disciplines. In the term transdisciplinary, the committee sees leveraging of existing concepts and approaches from multiple disciplines to derive new concepts and approaches, which in turn enable new ways to achieve and utilize understanding. Hence, transdisciplinary implies an integration-driven emergence of new disciplines, not just ad hoc collaborations." The report focused on the challenges of the US research sector, arguing that the "current organization of the research sector complicates communication and collaboration across disciplines." (*ARISE II*, p. xi)

Focusing on the physical sciences and engineering, and the life sciences and medicine, *ARISE II* takes a practice-based approach, arguing that incipient global challenges and new promising areas of research *require* collaboration and "deep conceptual and functional integration across scientific disciplines." The human genome project is held up as an example, as it required collaboration across multiple disciplines.

ARISE II argued that future grand challenges would require similar collaborations and transdisciplinary research, and furthermore, the US research sector was not organized effectively to enable this. For the purposes of this book's argument, transdisciplinarity *also* does much to undermine the basic/applied dichotomy. By their very nature transdisciplinary projects and collaborations cross boundaries and upend silos. The same is true of multidisciplinary work.

Linking the emerging research opportunities to emerging social challenges, such as climate change and fossil dependency, the report goes on to argue that the barriers to collaborative research lie in "dated and inflexible policies and organizational structures" and proposes a number of recommendations to bring about change.

The first goal of the ARISE II report was to promote transdisciplinarity, the second was to create a policy prescription aimed at taking a holistic approach to the entire research enterprise. It identified the need for greater collaboration and synergy across the entire spectrum of research activity, in academia, industry, and government research facilities.

> Creating an interdependent ecosystem requires incentives for basic and applied research, development, and deployment. Novel discoveries can emerge during the development process, and new technologies can arise out of basic research labs. The academic, government, and private sectors must develop an inclusive and adaptive environment that ensures that the unique objectives, skills, and points of view of the different sectors are integrated and optimally utilized. (*ARISE II*, p. xiv)

This report takes seriously the idea that the very structure of the US research enterprise needs to be remodeled. The concrete suggestions given for accomplishing these two major objectives are worth reviewing in some detail.

For the first goal the report recommends:

1. The fostering and development of a new multidisciplinary knowledge network.

2. An expanded educational program that would serve to model transdisciplinary approaches.
3. Expanded support for shared core research facilities.
4. Align appointment and promotion policies to support the new focus on transdisciplinarity.
5. Review administrative policies to streamline them to promote transdisciplinarity.

For the second goal it also recommends:

1. The establishment of grand challenges to motivate research cooperation among academia, industry, and government research.
2. The development and implementation of new models for research alliances between industry and the academy.
3. An enhancement to the permeability between industry and academia at all stages.
4. New priorities for technology transfer between industry and the academy.
5. The development of policies that highlight common areas of interest between industry and the academy and minimize areas of conflict.
6. The creation of mechanisms to increase coordination and cooperation among government agencies that support the PSE and the LSM, that is, the NSF and the NIH.

Taken as a whole, these prescriptions would address some of the criticism we have raised. For one, it is clear that the research enterprise imagined in *ARISE II* is multifaceted, complex, and contingent, requiring multiple kinds of expertise, and takes place through the involvement of various stakeholders—that is, academia, industry, and the government. The authors of *ARISE II* recognize the various feedback loops that the practice of research depends on and is sustained by. Even though it preserves the language of "basic" and "applied," its embrace of industrial research environments is a hallmark of this particular report and in this singular respect, *ARISE II* does much to restore the importance and relevance of industrial research and its contributions to science and technology in the United States.

The third and final policy prescription we will consider is *Restoring the Foundation: The Vital Role of Research in Preserving the American Dream*[10] published in September 2014. Cochaired by Norm Augustine and Neal Lane, the report is a comprehensive survey of the US research enterprise and contains an impressive number of concrete suggestions—most of which we agree with.[11] As the prescriptions enumerated in the *Restoring the Foundation (RtF)* report are quite substantive and specific, replicating them here would take up far too much space, so we will limit our comments to the areas where our critique diverges from the report.

Upon initial reading, it is clear that the RtF report is written in much the same vein as *Science, the Endless Frontier.* It calls for an intervention to the very bedrock of the American psyche—the American Dream. Both of these reports posit the argument that the American Dream is in jeopardy from globalization and from an increasing perception among the current generation of scientists and researchers that they confront a lack of opportunities. For the first time in quite a while, Americans are afraid that their children will not enjoy the same standard of living that they have enjoyed. It is against this backdrop that the twin forces of scientific and technological advancement are held up as the solution to preserve the American Dream and drive GDP growth.

The report argues that technological advancement is traceable to "research discovery"—effectively basic research—and *here is where we diverge* from the report. As we argued in previous chapters, research operates in a virtuous cycle. The entire reasoning of the discovery-invention cycle implies that each can inform the other. That means that technological advancements can lead to research discoveries and vice-versa. We want there to be no confusion: the linear model is incorrect.

The inductive steps of logic that the report undergoes to connect investments in "basic research" to American prosperity could be read as a more refined and subtle restatement of the linear model. In essence the report argues that "invest in basic research and you will get new products and technologies and increasing economic growth." While there is no doubt in our minds that the report authors have a much more complex view of the interaction of science, engineering, technology, and the economy, by not highlighting the bidirectional nature of science and technology and the catholic nature of research, the report could reinforce linear model thinking.

However, in spite of these points of departure from the report's argument, we agree wholeheartedly with its general thrust and endorse its broad claim about the need to provide increased funding for research as a proportion of GDP—with the critical caveat that discovery research and invention be acknowledged as two parts of the same coin, and that the increased funding should not impose limits on the activities of researchers in ways that limit the natural evolution of their work.

In conclusion, the discovery-invention cycle, in its acknowledgment of the multifaceted nature of research as resulting in both discovery and invention either separately or simultaneously, and by drawing the line at a different point, that is, between research and development, avoids these pitfalls. Much of the responsibility for the dilemma can be laid at the feet of the limited language of the basic/applied dichotomy and how difficult it is to express complex and contingent inter-relationships in the simplistic language of basic vs applied. This is one of the reasons why this book has advocated a shift in terminology toward the discovery and invention cycle, and provides the cognitive model that is necessary for such a shift to be broadly taken up by science and technology policy thought leaders, and, it is to be hoped, by lawmakers themselves.

9

MOVING FORWARD IN SCIENCE
AND TECHNOLOGY POLICY

· ·

HAVING INTRODUCED THE DISCOVERY-INVENTION cycle and contrasted it with the challenges and problems that have arisen from embracing the "basic"/"applied" framework, it is useful to return now to the scene with which we began this book. As a reminder at a congressional hearing, Secretary of Energy and distinguished researcher and Professor of MIT, Ernest Moniz was asked "Are the office of science's basic research programs a lower priority for this administration, when compared with these renewable programs?" Later Representative Hultgren (R-IL) stated that, "There's a problem that has been ongoing with this administration's choice to value applied R&D over basic scientific research."

As we describe in the first chapter, the problem here is not the constitutionally granted power of the purse that Congress should rightfully exercise, rather it is in the Congressional understanding of the research process as two separate and conflicting activities–"Basic Research" and "Applied Research." This is in direct conflict with the practical experiences of researchers as we have shown through multiple examples in this book. The difference between the congressional view of research and the practical understanding of research as a continuous cycling between discovery and invention helps us understand Secretary Moniz's response—"We think we have to work across the entire innovation chain."

142

In summary, there is an urgent pressing imperative to re-architect US national science and technology policy. We must move beyond the dichotomy of basic/applied and toward a synthesis of discovery and invention. We must recognize the historical nature of the basic vs. applied framework. It was a product of a particular time in which large personalities from the scientific and engineering communities were striving for dominance and influence over national policy. As we have described previously, the two world wars precipitated a shift in influence from the great industrialists (Edison and Bell) to the rising community of physicists (exemplified by physicists such as Millikan and Einstein). Bush wrote his influential report during a period where physicists dominated national science and technology policy, and the distinguished electrical engineer with savvy political instincts downplayed his own field of engineering, elevating science—even in the title of the report. If Bush had written his report a few decades before or even today, who is to say he may not have done the exact opposite?

The point then is the current basic/applied framework (upon which most of our science and technology policy is based) is the legacy of a political battle for influence—a poor foundation for a lasting and effectual framework for science and technology—not a reasoned reflection on the practical needs of researchers. What we have argued in this book, is that the basic vs. applied hierarchy needs to give way to a more nuanced understanding of the complex interplay of discovery and invention with no definitive hierarchical order. We believe that Bush's instincts were right when he called for a separation between activities aimed at unscheduled research (i.e., research) and activities directed toward the market (i.e., development). Our earlier review of the institutional structure of the best of the industrial laboratories highlights the need for some sort of separation. However, Bush drew the line incorrectly. We don't need to split the practice of research into two. Here are a few recommendations to move in the right direction of increased research productivity.

We Need New Language for S&T Policy

Though the linear model has been successfully undermined by social scientists (science and technology studies and economics in particular) as we demonstrate in this book, this rich work has not resulted in any meaningful change of the use and understanding of the terms "basic research" and "applied research." These linear model concepts are still alive and well. STS has clearly shown the social and political nature of research in science and technology (Bloor 1976; Latour 1987; Pielke Jr. 2012; Winner 1980; Jasanoff 2004) and economics has proposed more complex understandings of innovation and research (Nelson 1993).

Indeed the most significant critique from the social sciences that has achieved some level of acceptance in policy circles has been Donald Stokes' Pasteur's Quadrant, which introduced qualifiers of "pure" vs. "use-inspired" to the basic/applied dichotomy. This only served to reinforce the division of research by placing "the quest for understanding" in a different axis from "considerations of use." Also Stokes' preserved the terms "basic" and "applied." As long as the distinction of research into a "basic" type vs. an "applied" type still holds, the linear model will remain a phoenix, constantly beaten down, but destined to rise again.

It is also clear that the various blue ribbon reports and panels we review in the previous chapter are sensitive to the need to rearchitect national science and technology policy. However, until the intellectual basis of the current US science and technology policy changes, no true and lasting reformation is possible. This is why we believe that as a first step, it is necessary to change the language used to describe research. It is necessary to move away from outdated models that this language keeps alive and use new terminology as a cognitive aid to bootstrap a new kind of thinking into the policy discourse. Nomenclature is important. We should immediately drop the use of the terms basic research and applied research and instead talk about "research" with the understanding that it encompasses both invention and discovery.

Research is an unscheduled quest for new knowledge and the creation of new inventions, whose outcome cannot be predicted in advance, and in which both science *and* engineering are essential ingredients. The boundary should be drawn at development, not applied research, which should disappear as a term. For us, development is a scheduled activity with a well-defined outcome in a specified time frame, aimed at the marketplace. It does need to be distinguished from research, lest short-term imperatives override the need for un-scheduled creative research. Even in this, the separation should be managed carefully and must be one of insulation, not isolation, allow-ing researchers the individual freedom to move across the boundary.

One of the major advantages of adopting new language in science and technology policy is that it would necessitate a reconsideration of the currently existing problematic categories and definitions which have resulted from an embrace of "basic" and "applied" research. For example, as we document in Chapter 4, even in an organization like the Department of Defense, (which in many ways is a natural ally of the discovery and invention cycle and has in various reports[1] and initiatives recognized the need for a broad definition of research) is still limited by inflexible research categories can severely constrain the ability of researchers to take their work to its natural conclusion.

We Need to Restructure Research Environments

Second, we must restructure our research environments. Research is an activity that flourishes only in particular, specific environments. As we have shown in previous chapters, the institutional elements necessary for the flourishing of research include: a well-defined vi-sion and mission; a transparent, meritocratic, and entrepreneurial culture; stable and flexible funding at all appropriate scales; insu-lation but not isolation; and a leadership cadre that is technically distinguished and capable of exercising exquisite judgment in main-taining the fine balance between focus and freedom that is necessary to incentivize and inspire researchers. A successful and visionary research institution also requires an administrative staff who see their primary responsibility as one of supporting and facilitating the research effort—all these elements need to be attended to at every scale of the institution—from individuals to departmental units, to the entirety of the institution. These institutional elements must be replicated in all fractal dimensions from the smallest to the largest unit.[2] Many of the great industrial research laboratories, especially Bell Labs, exemplified these characteristics at various levels in the structure of the institution.

The cases we have considered in this book demonstrate that variations upon this theme can be successfully implemented in modern research institutions—including our research intensive universities. A prime example is found at the University of California at Santa Barbara, which achieved national prominence in the physical sciences and engineering over the period of only a few decades. This was done through visionary leadership and the strategic recruitment of relevant individuals and necessary resources, with the goal of fostering a cross-disciplinary research culture and establishing a critical mass in selected areas. Janelia Research Campus is another example from a different field of endeavor. The Janelia laboratory has brought together diverse fields of the life sciences, computation and physical sciences, and engineering in an effort to tackle the brain—an identified grand challenge problem in neuroscience and biology.

The Advancing Research in Science and Engineering II (*ARISE II*) report, discussed in the previous chapter, has a number of examples of innovative institutional designs, such as the Energy Biosciences Institute and the Structural Genomic Consortium.[3] We will not discuss these here. However, there are a few additional institutions that are worth describing in addition to those discussed above—particularly the NSF and its efforts in this area.

In the 1980s the National Science Foundation under the directorship of Erich Bloch (former vice president of IBM), fostered the creation of the Science and Technology Centers (STCs) and Engineering Research Centers (ERCs)[4] at universities in a clear attempt to bridge the basic applied dichotomy. Much later, the US Department of Energy, a mission-oriented agency (unlike NSF) saw the light under Secretary Chu's leadership, when it created the first Energy Frontier Research Centers (EFRCs) in 2009. These centers have as their stated aim, the need to tackle the hard problems in energy by bringing together researchers from universities and the national laboratories in a bid to accomplish this goal. Also promising are the larger efforts involving industry, the national laboratories, and universities. These efforts are aimed at creating alternative energy technologies and are organized as the Energy Innovation Hubs (EIH). These so-called "mini-Bell" labs are still in their infancy and will need to be monitored and evaluated to effect enduring changes in the system. These laudable efforts are held back by the bureaucratic resistance that the basic/applied framework generates as documented in Chapter 1.

Turning for a moment to the national laboratories, it is clear that they have the scale and breadth of mission (for example, in energy and defense) to be directly comparable to Bell Labs. They could accurately be described as a system, spanning all phases of research activity including the early stages of creating new technologies. The national labs also have a well-defined mission of technology translation to industry. However, when one considers the implications of the discovery-invention cycle as it applies to the national laboratories, it is immediately clear that at least in the area of energy, the required changes are sufficiently radical enough to necessitate a revisiting of the structure of the Department of Energy's oversight of the labs, and move it toward something more like the original government owned/contract operated (GOCO) concept. This should replace the current hybrid civil-service type of institutional arrangement that the original GOCO system has degenerated into.[5]

Most of our arguments in this book could be read as being most applicable to two major mission oriented agencies—the Department of Energy (DOE) and the National Institutes of Health (NIH)—which fund, along with the National Science Foundation, a significant portion of university-based research. As we document in the first chapter, these two particular agencies (NIH and DOE) face significant difficulty in bridging the basic-applied dichotomy to address grand challenge problems. This is a problem that other mission-oriented agencies do not have as they are *performing agencies*. That is, the Department of Defense and the National Aeronautics and Space Agency are more engaged with the discovery-invention cycle, because in large measure the government is both the *funder* and the *client*. This situation is very similar to that at AT&T where Bell Laboratories engaged in both discovery and invention tied to AT&T specific mission with a well-defined client (AT&T/Western Electric). Contrast this with the DOE and the NIH which even though they have a clear mission, who their client is, is not clear. One way to resolve this would be through closer ties to industry (with regard to the DOE[6]) and clinical research (in the case of the NIH).

In summary, reworking these national treasures to allow for greater research interactivity will require cultural shifts along the lines of discovery and invention and away from hierarchical basic/applied descriptors. This reworking should be different for each national laboratory. The leadership challenges at the laboratories (as opposed to mere technical administration) will also require a shift away from a culture of bureaucratic list checking toward a culture of intellectual risk taking, which can only be accomplished by empowering the laboratory directors with the freedom necessary to accomplish the technical mission.

A New Analytical Framework for S&T Policy

We feel that the time has come to seriously rethink the intellectual basis of our science and technology policy. The basic/applied framework no longer serves the nation well. It should be relegated to the dustbin of history. This is particularly important in light of the current cultural moment where engineering, information technology and the computational sciences are in the ascendancy, and are being viewed in a more balanced way in the S&T ecosystem. Reductionism has been supplanted by integrative and systems thinking especially in addressing the grand challenges facing society. For example, information and communication technologies are becoming more prominent in tackling global health challenges—especially in the most resource deprived parts of the world. Computation is playing an increasing role in biology and medicine. Big data holds the promise of potentially creative solutions to some of the broad problems facing the planet in transportation, climate change, and conservation.

Unlike previous decades in science and technology policy, which were marked by a contentious shift of influence from the great industrialists to the great physicists after the war, it is clear that at the present time, engineering no longer needs to be in competition with science. This is a time when we venerate great theoretical physicists such as Peter Higgs and Stephen Hawking, *and* also great engineers and innovators such as Bill Gates, Steve Jobs, Elon Musk, Sergey Brin, and Larry Page. This is the age of engineering as evidenced by the establishment of schools of engineering at various elite institutions (for example, Harvard and Yale), and new types of research institutions that bridge the life sciences with engineering and the physical sciences (Janelia Research Campus is a prime example of this).

What we need to sustain and extend this shift, is a reframing of the intellectual basis of the relationship between science, technology, and engineering, placing each on an equal footing with respect for both the abstract and the concrete. It is time for the old hierarchies to be put aside and US science and technology policy needs to accept as a foundational premise, the equivalent value of abstraction and experimentation. It is time for US science and technology policy to drop the basic/applied framework and embrace a framework for research that is aligned with the need to address complex twenty-first century problems. It is time for our policy frameworks to enhance, not create difficulties for the practice of research. The discovery-invention cycle is our suggestion for this new framework.

ABBREVIATIONS

Advanced Research Projects Agency-Energy (ARPA- E)
Advancing Research in Science and Engineering (ARISE)
Air Force Office of Scientific Research (AFOSR)
American Association for the Advancement of Science (AAAS)
American Telephone and Telegraph Co. (AT&T)
Defense Advanced Research Projects Agency (DARPA)
Department of Defense (DoD)
Department of Energy (DOE)
Energy Frontier Research Center (EFRC)
Energy Innovation Hub (EIH)
Engineering Research Center (ERC)
General Electric (GE)
International Business Machines (IBM)
National Academy of Engineering (NAE)
National Academy of Sciences (NAS)
National Aeronautics and Space Administration (NASA)
National Institute of Health (NIH)
National Institute of Standards and Technology (NIST)
National Research Council (NRC)
National Science Foundation (NSF)
Office of Naval Research (ONR)
Office of Scientific Research and Development (OSRD)
Research and Development (R&D)
Science and Technology (S&T)
Science, Technology Center (STC)
University of California Santa Barbara (UCSB)

NOTES

CHAPTER 1 · *Breaking Barriers, Building Bridges*

1. See Conover, "Moniz Defends DOE Budget at House Hearings," 2015.
2. See Cho, U.S. House Panel Would Slash Department of Energy's Applied Research," 2015.
3. William Shockley, "Transistor Technology Evokes New Physics," 1956b.
4. See Roosevelt, 1944.
5. Yes, Bardeen was an engineer! He graduated with a BS in electrical engineering in 1928 and worked for two years as a research assistant in electrical engineering. See Nobel Lectures, 1964.
6. https: // www.whitehouse.gov / share / brain-initiative.
7. Insel et al., "The NIH BRAIN Initiative," 2013.
8. While few members of the committee have physical science backgrounds, all have established careers in neuroscience.
9. http: // braininitiative.nih.gov / 2025 / index.htm.
10. See Narayanamurti et al., "ARISE II," 2013.
11. See Gieryn, "Boundary-work and the Demarcation of Science from Non-science," 1983, for a discussion of boundary work as a practical and political process.
12. Pielke, "Basic Research as a Political Symbol," 2012, undertakes a clear discussion of the emergence of basic research as a political symbol and the implications for trying to bring about change in US science and technology Policy.

CHAPTER 2 · *Boundaries in Science and Engineering Research*

1. Ringertz, "Alfred Nobel—His Life and Work," n.d.
2. Lundström, "Alfred Nobel's Dynamite Companies," 2003.
3. "Full Text of Alfred Nobel's Will," n.d.
4. "Thomas Edison and Menlo Park," n.d.
5. One only need review the 2012 State of the Union (SOTU), or any of the last few SOTUs for examples of the persistence of the "basic" vs. "applied" frame. There are of course, exceptions to this observation. An especially notable one is, *Rising above the Gathering Storm*.

6. Harvard, n.d.

7. "Alumni Petition Opposing MIT-Harvard Merger, 1904–05: Exhibits: Institute Archives and Special Collections: MIT," 2005.

8. Bill Gates was in the class of 1977. He received an honorary doctorate degree (LL.D) in 2007. Steve Ballmer graduated magna cum laude from Harvard in 1977.

9. In a previous published paper, we argued that the terms "basic" and "applied" and the dichotomy they represent have outlived their usefulness and now act as barriers to organizing effective research environments.

10. Schaffer, "The Laird of Physics," 2011.

11. Geselowitz, "Did You Know? Someone Else Wrote Maxwell's Equations," 2013. Geselowitz discusses Heaviside's contributions to electrical engineering and his mastery of applied mathematics and influence on the modern form of Maxwell's equations in the IEEE piece. Also, Knoll points out that the mathematicians were suspicious of Heaviside's mathematical methods even as he used them to clarify Maxwell's work.

12. In a classic case of coinvention, the German Heinrich Hertz also described the four equations simultaneously. Indeed for a time they were known not as Maxwell's equations, but as the Hertz-Heaviside equations.

CHAPTER 3 · *The Basic/Applied Dichotomy: The Inadequacy of the Linear Model*

1. Bush, "Science, the Endless Frontier," 1945.

2. Using the crude estimate of word counts, Bush uses the term science about 120 times, but engineering only appears about four times in the entire report, a ratio of 30:1.

3. Baxter, *Scientists against Time*, 1946. On page 142, Baxter (1893–1975) describes this as "the most valuable cargo ever brought to our shores."

4. Shapin, *The Scientific Life: A Moral History of a Late Modern Vocation*, 2008; Stokes, *Pasteur's Quadrant: Basic Science and Technological Innovation*, 1997.

5. Godin, "The Linear Model of Innovation," 2006.

6. Ibid., p. 640.

7. Roosevelt, Papers as President, 1933. For example in President Roosevelt's letter, the idea of diffusing the results of scientific knowledge in order to stimulate new enterprises was clearly articulated.

8. Layton, "Mirror-Image Twins: The Communities of Science and Technology in 19th-Century America," 1971.

9. Layton Jr, "Technology as Knowledge," 1974.

10. Buvet et al., "Living Systems as Energy Converters." 2013. See George Porter's contribution to the edited volume for more information on this statement.

11. Nelson, *National Innovation Systems*, 1993, p. 4.

12. Stokes, *Pasteur's Quadrant: Basic Science and Technological Innovation* 1997b.

13. Ibid., 88.

14. Narayanamurti et al., "ARISE II," 2013. Narayanamurti was a cochair of this report. *ARISE II: Unleashing America's Research and Innovation Enterprise*, (Cambridge, MA: American Academy of Arts and Sciences).

15. Ibid., p. XX.

16. Casimir, *Haphazard Reality: Half a Century of Science*, 1983.

17. For an example of a positive engagement of Casimir's work and its application to research environments see Tsao et al., "Galileo's Stream: A Framework for Understanding Knowledge Production." 2008.

18. Layton, 1971.

CHAPTER 4 · *The Origins of the "Basic" and "Applied" Descriptors*

1. Kline, "Constructing Technology as Applied Science," 1995, p. 196.

2. Rowland, "A Plea for Pure Science," 1883

3. Ibid.

4. Kline, "Constructing Technology as Applied Science," 1995 p. 209.

5. For more on the Weber on bureaucracy see Weber, *From Max Weber: Essays in Sociology*, 2009.

6. Narayanamurti et al., "ARISE II: Unleashing America's Research and Innovation Enterprise," 2013. Much of the argument of the paper is replicated in Chapter 5.

7. Personal communication, see Appendix 1 for full details.

8. Whitney, "Research as a National Duty," 1916.

9. Marshall, "Edison's Plan for Preparedness," 1915.

10. Kevles, *The Physicists: The History of a Scientific Community in Modern America*, 1995, p. 109.

11. Ibid.

12. Ibid., p. 138.

13. See Godin, *Science, Technology, and Human Values*, 2006. It contains a discussion of how the linear model of innovation evolved.

14. See Pielke Jr., "Basic Research as a Political Symbol," 2012 for a thorough discussion of Schumpeter and Solow and their use in linear model arguments.

15. See Cech et al., "ARISE I—Advancing Research in Science and Engineering: Investing in Early-Career Scientists and High-Risk, High-Reward Research," 2008. The report contains a discussion of the forced separation of the life sciences from the practice of medicine.

CHAPTER 5 · *The Discovery-Invention Cycle*

1. A version of this argument was previously published in article, Narayanamurti et al., RIP: The Basic/Applied Research Dichotomy," 2013.

2. By analyzing the processes of invention and discovery in physics Nobel Prizes, long considered the most mathematical and analytically precise of the sciences, we are continuing the fine tradition of undermining the specious claim that science precedes technology.

3. For example, the many physics and chemistry Nobel Prizes concerned with nuclear magnetic resonance and imaging.

4. Interestingly, in his lecture, Shockley discusses the usefulness of the classifying terms "pure," "applied," "fundamental," and "basic" as they are used to describe research in science and technology. For Shockley, these terms are too

often used in a derogatory sense to elevate research that is driven by a motivation of "aesthetic satisfaction" over research driven by a desire to improve a process. See W. Shockley, "Transistor Technology Evokes New Physics,"1956b, p. 345.

5. His postdoc, brother-in-law and collaborator at Bell Labs Arthur Schawlow (1981 Nobel Laureate) worked with Townes on developing the laser, which was circulated in preprints at both institutions and patented by Bell Labs after Townes's insistence, despite the reluctance of Bell Labs attorneys who did not see worth in the patent.

6. See discussion of the NMR cycle below.

7. A well-known example of a mission that led to both discovery and invention from the field of computer science and electrical engineering that we have not discussed in any sort of detail is the development of the early Internet. Research on the early Internet was driven by funding from the Defense Advanced Research Projects Agency (DARPA) with the stated goal of creating a robust communication network that would be resilient to attack and disruption. This led to further research and the creation of new fields in computer science and can also be directly linked to the birth of the computational sciences and engineering in fields as diverse as computational biology, robotics, and the Internet.

CHAPTER 6 · *Bell Labs and the Importance of Institutional Culture*

1. Bimberg, "A Tribute to Zhores Ivanovitch Alferov, a Pioneer Who Changed Our Way of Daily Life," 2011.

2. Like Bush, *Pasteur's Quadrant* [Stokes 1997] framework still retains the basic/applied distinction and is not very useful here.

3. Gertner, *The Idea Factory: Bell Labs and the Great Age of American Innovation*, 2012a.

4. Noll and Geselowitz, *Bell Labs Memoirs: Voices of Innovation*, 2011.

5. Gertner, *The Idea Factory: Bell Labs and the Great Age of American Innovation*, 2012a.

6. Galambos, 1992; Gertner, *The Idea Factory: Bell Labs and the Great Age of American Innovation*, 2012a.

7. The Audion came to be known as the vacuum tube.

8. Gertner, *The Idea Factory: Bell Labs and the Great Age of American Innovation*, 2012a.

9. Bown, "Vitality of a Research Institution and How to Maintain It," 1953.

10. Quoted in Gertner, "True Innovation," 2012b.

11. William Shockley, "Transistor Technology Evokes New Physics," 1956b.

12. .Thierer, "Unnatural Monopoly: Critical Moments in the Development of the Bell System Monopoly," 1994.

13. Kelly, *The First Five Years of the Transistor*, 1953.

14. A note on methods: We interviewed a large number of former Bell Labs members of technical staff who worked at the labs during the 1960s, 1970s, and 1980s. Beyond the requirement that their service at Bell Labs had to fall sometime within those three decades, we also attempted to include as much diversity of experience, laboratory affiliation, and disciplinary focus in our interview pool as we could. A full list of all interviewees is available in the Appendix.

The interviews were semi-structured in nature and we later coded the interviews, leading to the analysis above and the identification of elements of research culture.

15. Noll and Geselowitz, *Bell Labs Memoirs: Voices of Innovation* 2011.

16. Tsao et al., "Art and Science of Science and Technology: Proceedings of the Forum and Roundtable," 2013.

17. Noll and Geselowitz, *Bell Labs Memoirs: Voices of Innovation*, 2011.

18. Tsao et al., "Art and Science of Science and Technology: Proceedings of the Forum and Roundtable." 2013.

19. Bill Brinkman's distinguished career includes the thirty-five years he spent at Bell Labs, a stint as vice president of research at Sandia National Laboratories, and then he joined the Department of Energy as the head of the Office of Science in 2009.

20. Narayanamurti was an MTS in the Solid State and Quantum Physics Research Department – Department 1116, before becoming the head of Department 1154—the Semiconductor Electronics Research Department in 1976. In 1981, he became the director of the Solid State Electronics Research Laboratory (115). In 1987, he became vice president of research at Sandia National Laboratories which at that time was operated under contract by AT&T/Western Electric.

21. Stewart, "At Google, a Place to Work and Play," 2013.

CHAPTER 7 · *Designing Radically Innovative Research Institutions*

1. See *Advancing Research in Science and Engineering II* and *Rising above the Gathering Storm.*

2. The acknowledgements section in this volume contains a comprehensive list of interviewees from both institutions.

3. http:// dealbook.nytimes.com / 2014 / 10 / 24 / a-billionaires-65-million-gift-to-theoretical-physics / ?_php=true&_type=blogs&_r=0.

4. Kinney, A. L. "National Scientific Facilities and Their Science Impact on Nonbiomedical Research," 2007.

5. http:// www.forbes.com / sites / liyanchen / 2014 / 07 / 30 / start-up-schools-americas-most-entrepreneurial-universities /.

6. Howard Hughes Medical Institute, "Janelia Farm Research Campus: Report on Program Development," 2003.

7. See http://www.biotechniques.com/news/Fluorescent-Protein-Biosensors-Arent-Bulletproof-Yet/biotechniques-326771.html.html for a discussion of the biosensors work at Janelia.

CHAPTER 8 · *The Need for a Radical Reformulation of S&T Policy*

1. Augustine et al., *Rising above the Gathering Storm*, 2005.

2. Augustine et al., *Rising above the Gathering Storm, Revisited: Rapidly Approaching Category 5*, 2010.

3. De Gennes and Badoz, *Fragile Objects: Soft Matter, Hard Science, and the Thrill of Discovery*, 1996.

4. See Comte, A. *The Course of Positive Philosophy*, 1876.

5. See Whitesides, G.M. and Deutch, J., 2011. Let's Get Practical. *Nature*, 469(7328), pp. 21–22. Also, Whitesides, G.M., 2015. Reinventing Chemistry. *Angewandte Chemie* [International Edition], 54(11), pp. 3196–3209.

6. See Mervis, "Contentious Markup Expected Today as House Science Panel Takes Up COMPETES Bill," 2015.

7. Cech et al., *ARISE I*, 2008.

8. Narayanamurti et al., *ARISE II: Unleashing America's Research and Innovation Enterprise*, 2013.

9. For more on the issue of convergence of engineering and the life sciences, see Sharp and Langer, "Promoting Convergence in Biomedical Science," 2011.

10. Augustine et al., "Restoring the Foundation: The Vital Role of Research in Preserving the American Dream, New Models for U.S. Science and Technology Policy," 2014.

11. Venkatesh Narayanamurti was a member of the report committee and one of the authors of this book.

CHAPTER 9 · *Moving Forward in Science and Technology Policy*

1. For example, see the 2012 report of the Defense Science Board Task force on Basic Research.

2. For more on fractal replication of research culture, see Narayanamurti et al., "Transforming Energy Innovation," 2009.

3. For more on the Energy Biosciences Institute and the Structural Genomic Consortium see Anadon, Bunn, and Narayanamurti, *Transforming U.S. Energy Innovation*, 2014; Narayanamurti et al., *ARISE II*, 2008.

4. See Currall, *Organized Innovation: A Blueprint for Renewing America's Prosperity*, 2014) for a discussion of the NSF ERC program and its return to the US economy.

5. For more on the national laboratories, see Narayanamurti, et al., "Institutions for Energy Innovation: A Transformational Challenge, 2009." Also, Anadon, Bunn, and Narayanamurti, *Transforming U.S. Energy Innovation*, 2014.

6. See also the findings of the Commission to Review the Effectiveness of the National Energy Laboratories (CRENEL) which published a draft report in September of 2015.

BIBLIOGRAPHY

Alumni Petition Opposing MIT-Harvard Merger, 1904–05: Exhibits: Institute Archives and Special Collections, 2005. Cambridge: MIT Library. Retrieved from https: // libraries.mit.edu / archives / exhibits / harvard-mit /.

Anadon, L. D., M. Bunn, and V. Narayanamurti. 2014. *Transforming U.S. Energy Innovation*. New York: Cambridge University Press.

Augustine, N. R., C. Barrett, G. Cassel, N. Grasmick, C. Holliday, S. A. Jackson, SA. K. Jones, R. Levin, C. D. Mote, and C. Murray. 2005. *Rising* above *the Gathering Storm*. Washington, DC: National Academies Press.

Augustine, N. R., C. Barrett G. Cassell, N. Grasmick, C. Holliday, S. A. Jackson, A. K. Jones, R. Levin, C. D. Mote, and C. Murray. 2010. *Rising above the Gathering Storm, Revisited: Rapidly Approaching Category 5*. Washington, DC: National Academies Press.

Augustine, N. R., N. Lane, N. Andrews, J. E. Bryson, T. R. Cech, S. Chu . . . E. Zerhouni. 2014. *Restoring the Foundation: The Vital Role of Research in Preserving the American Dream, New Models for U.S. Science and Technology Policy*. Cambridge, MA: American Academy of Arts & Sciences. Retrieved from https://www.amacad.org/content/Research/researchproject.aspx?d=1276.

Baker, W. O., 1965. "Engineering and Science: A Sum and Not a Difference," *Listen to Leaders in Engineering*, ed. Albert Love and James Saxon Childers, (pp. 297–309). Atlanta: Tupper and Love.

Baxter, J. P., 1946. *Scientists against Time*. Boston: Little, Brown and Company.

Bimberg, D., 2011. "A Tribute to Zhores Ivanovitch Alferov, a Pioneer Who Changed Our Way of Daily Life." *Semiconductor Science and Technology* 26. doi: 10.1088/0268–1242/26/1/010301

Bloor, D., 1976. *Knowledge and Social Imagery*. London: Routledge and Kegan Paul.

Bown, R., 1953. "Vitality of a Research Institution and How to Maintain It." *Bell System Technical Journal, Publications Monograph* 2207: 1953.

Bush, V., 1945. "Science, 'The Endless Frontier.'" *Transactions of the Kansas Academy of Science* 1903: 231–264.

Buvet, R., M. J. Allen, and J.-P. Massué, 2013. "Living Systems as Energy Converters." Proceedings of the European Conference on Living Systems as Energy Converters, Organized Under the Auspices of the Parliamentary Assembly of the Council of Europe in Collaboration with the Commission of European Communities, Pont-À-Mousson, France, September 24, 2013.

Casimir, H. B. G., 1983. *Haphazard Reality: Half a Century of Science*, Alfred P. Sloan Foundation Series. New York: Harper & Row.

Cech, T. R., D. Baltimore, S. Chu, F. Cordova, R., Everhart, T., Freeman, S. . . . Zoghbi, H. 2008. "ARISE I—Advancing Research in Science and Engineering: Investing in Early-Career Scientists and High-Risk, High-Reward Research," Report prepared for Advancing Research in Science and Engineering. Cambridge, MA: American Academy of Arts and Sciences. Retrieved from https://www.amacad.org/multimedia/pdfs/publications/books/ariseReport.pdf

Cho, A., 2015, April 22. "U.S. House Panel Would Slash Department of Energy's Applied Research." *Science*. doi:10.1126 / science.aab2535

Comte, A. 1876. *The Course of Positive Philosophy*. London: George Bell and Sons.

Conover, E., 2015, February 27. "Moniz Defends DOE Budget at House Hearings". *Science*. doi:10.1126 / science.aaa7923

Currall, S. C., E. Frauenheim, S. J. Perry, and E. M. Hunter. 2014. *Organized Innovation: A Blueprint for Renewing America's Prosperity*. New York: Oxford University Press.

De Gennes, P. -G., and J. Badoz, 1996. *Fragile Objects: Soft Matter, Hard Science, and the Thrill of Discovery*. New York: Springer.

Galambos, L., 1992. "Theodore N. Vail and the Role of Innovation in the Modern Bell System." *Business History Review* 66: 95–126.

Gertner, J., 2012a. *The Idea Factory: Bell Labs and the Great Age of American Innovation*. New York: Penguin Press.

Gertner, J., 2012b. "True Innovation," *New York Times*, Sunday Review, February 25.

Geselowitz, M., 2013. "Did You Know? Someone Else Wrote Maxwell's Equations." Retrieved from http://theinstitute.ieee.org/technology-focus/technology-history/did-you-know-someone-else-wrote-maxwells-equations

Gieryn, T. F., 1983. "Boundary-Work and the Demarcation of Science from Non-Science: Strains and Interests in Professional Ideologies of Scientists," *American Sociological Review* 48: 781–795.

Godin, B., 2006. "The Linear Model of Innovation: The Historical Construction of an Analytical Framework," *Science, Technology, and Human Values* 31: 639–667. doi:10.1177 / 0162243906291865

Harvard John A. Paulson School of Engineering and Applied Sciences. n.d. "Gordon McKay." Retrieved from www.seas.harvard.edu / about-seas / history-seas / founding-early-years / gordon-mckay

Howard Hughes Medical Institute, 2003. "Janelia Farm Research Campus: Report on Program Development." Chevy Chase, MD: Howard Hughes Medical Institute.

Insel, T. R., S. C. Landis, S. C., and F. S. Collins, 2013. "The NIH BRAIN Initiative." *Science* 340, 687–688. doi:10.1126 / science.1239276

Jasanoff, S. (Ed.). 2004. *States of Knowledge: The Co-Production of Science and the Social Order*. New York: Routledge.

Kelly, M. J., 1953. *The First Five Years of the Transistor*. New York: American Telephone and Telegraph Company.

Kevles, D. J., 1995. *The Physicists: The History of a Scientific Community in Modern America*, rev. ed. Cambridge: Harvard University Press.

Kinney, A. L. "National Scientific Facilities and Their Science Impact on Nonbiomedical Research." Proceedings of the National Academy of Sciences 104, no. 46 (November 13, 2007): 17943–47. doi:10.1073 / pnas.0704416104

Kline, R., 1995. "Construing 'technology' as 'applied science': Public Rhetoric of Scientists and Engineers in the United States, 1880–1945." *Isis* 86: 194–221.

"Latour, B., 1987. *Science in Action: How to Follow Scientists and Engineers through Society*. Milton Keynes and Philadelphia: Open University Press.

Layton, E., 1971. "Mirror-Image Twins: The Communities of Science and Technology in 19th-Century America," *Technology and Culture*, 562–580.

Layton Jr., E. T., 1974. "Technology as Knowledge," *Technology and Culture* 15(1): 31–41.

Lundström, R., 2003. Alfred Nobel's Dynamite Companies. Retrieved from https://www.nobelprize.org/alfred_nobel/biographical/articles/lundstrom/

Marshall, E., 1915. "Edison's Plan for Preparedness," the *New York Times*, May 30, SM6.

Mervis, J., 2015, April 22. "Contentious Markup Expected Today as House Science Panel Takes Up COMPETES Bill." *Science*. doi:10.1126 / science.aab2529.

Narayanamurti, V., L. D. Anadon, and A. D. Sagar, 2009. "Institutions for Energy Innovation: A Transformational Challenge." Paper prepared for the Energy Technology Innovation Policy research group, Belfer Center for Science and International Affairs, Harvard Kennedy School, September.

Narayanamurti, V., L. D. Anadon, and A. D. Sagar, 2009. "Transforming Energy Innovation." *Issues in Science and Technology* 26, 57–64.

Narayanamurti, V., T. Odumosu, and L. Vinsel, 2013. "RIP: The Basic/Applied Research Dichotomy," *Issues in Science and Technology* 29, 31–36.

Narayanamurti, V., K. Yamamoto, N. Andrews, D. Ausiello, L. Bacow, M. Beasley, E. J. J. Benz, D. Botstein, et al., 2013. *ARISE II: Unleashing America's Research and Innovation Enterprise*. Cambridge, MA: American Academy of Arts and Sciences.

Nelson, R. R., 1993. *National Innovation Systems: A Comparative Analysis*. New York: Oxford University Press.

Nobel Lectures, 1964. Physics 1942–1962. Amsterdam: Elsevier–Retrieved from https://scholar.google.com/scholar?q=nobel+lectures&btnG=&hl=en&as_sdt=0%2C22

Nobelprize.org. n.d. Alfred Nobel's Will. Retrieved from http:// www.nobelprize.org / alfred_nobel / will /

Noll, A. M., Geselowitz, M., 2011. *Bell Labs Memoirs: Voices of Innovation*. Lexington, KY: CreateSpace Independent Publishing Platform.

Pielke Jr, R., 2012. "Basic Research as a Political Symbol." *Minerva* 50, 339–361.
Ringertz, N., n.d. "Alfred Nobel—His Life and Work." Retrieved from http: // www. nobelprize.org / alfred_nobel / biographical / articles / life-work /
Roosevelt, F. D., 1933. Papers as President. The President's Secretary's File. File Box 2. Retrieved from http://www.fdrlibrary.marist.edu/archives/collections/franklin/?p=collections/findingaid&id=502
Roosevelt, F. D., 1944. President Roosevelt's Letter to Vannevar Bush, November 24. Retrieved from http://scarc.library.oregonstate.edu/coll/pauling/war/corr/sci13.006.4-roosevelt-bush-19441117.html
Rowland, H. A., 1883. "A Plea for Pure Science." *Journal of the Franklin Institute* 116: 279–299.
Schaffer, S., 2011. "The Laird of Physics." *Nature* 471: 289–291. doi:10.1038 / 471289a
Shapin, S., 2008. *The Scientific Life: A Moral History of a Late Modern Vocation.* Chicago: University of Chicago Press.
Sharp, P. A., and Langer, R., 2011. "Promoting Convergence in Biomedical Science." *Science* 333: 527.
Shockley W. l956a. "Banquet Speech." Retrieved from www.nobelprize.org/nobel_prizes/physics/laureates/1956/shockley-speech.html
Shockley, W., 1956b. "Transistor Technology Evokes New Physics," In Nobel Lectures, Physics 1942–1962 (Elsevier, Amsterdam, 1964) 344–374.
Shockley, W., 1956b. The Nobel Prize in physics Speech presented at the Nobel Banquet in Stockholm, December 10,
Snow, C. P., 2012. *The Two Cultures.* New York: Cambridge University Press.
Stewart, J. B., 2013. "At Google, a Place to Work and Play." *New York Times,* March 15. Retrieved from http://www.nytimes.com/2013/03/16/business/at-google-a-place-to-work-and-play.html
Stokes, D. E., 1997. *Pasteur's Quadrant: Basic Science and Technological Innovation.* Washington, DC: Brookings Institution Press.
Thierer, A. D., 1994. "Unnatural Monopoly: Critical Moments in the Development of the Bell System Monopoly" *Cato Journal* 14, 267.
Thomas Edison and Menlo Park, n.d. Retrieved from http: // www.menloparkmuseum.org / thomas-edison-and-menlo-park
Tsao, J. Y., K. W. Boyack, M. E. Coltrin, J. G. Turnley, and W. B. Gauster. 2008. "Galileo's stream: A Framework for Understanding Knowledge Production. *Research Policy* 37(2): 330–352.
Tsao, J. Y., G. R. Emmanuel, T. Odumosu, A. R. Silva, V. Narayanamurti, G. J. Feist, G. W. Crabtree, C. M. Johnson, J. I. Lane, L. McNamara, S. T. Picraux, R. K. Sawyer, R. P. Schneider, C. D. Schunn, and R. Sun, 2013. "Art and Science of Science and Technology: Proceedings of the Forum and Roundtable." Presented at the June 5–7, 2013, Sandia National Laboratories, Albuquerque, New Mexico, Science, Technology, and Public Policy Program, Belfer Center for Science and International Affairs, Harvard Kennedy School.
Weber, M., 2009. *From Max Weber: Essays in Sociology.* Philadelphia: Routledge.
Whitney, W. R., 1916. "Research as a National Duty." *Science* 43: 629–637.
Winner, L., 1980. "Do Artifacts Have Politics?" *Daedalus* 109, 121.

ACKNOWLEDGMENTS

THE IDEAS AND ARGUMENTS in this book were painstakingly fleshed out during long hours of discussion and debate between the authors. We worked over Skype at Harvard University and at the University of Virginia, and we are grateful to both institutions for their material support. This book is based on years of personal experience and reflection and is influenced by literature from the fields of history of technology, the history of science, science and technology studies, electrical engineering, condensed matter physics, policy analysis, material science and engineering, public policy, and by various blue ribbon panel reports. Colleagues in these fields who have influenced our thinking are too numerous to name here. However, some essential individual contributions are particularly noteworthy and deserve to be highlighted.

This book grew out of the enthusiastic responses we received on a paper we published in *Issues in Science and Technology* with Lee Vinsel, and from Venky's early reflections on the genesis of the Rowland Institute at Harvard, an organization modeled in some parts on Bell Labs. It would not have been possible without initial comments from Graham Allison, John Deutch, Neal Lane, Arun Majumdar, C. Daniel Mote, Cherry Murray, Roger Pielke, Jr., Lyle Schwartz, and Jeff Tsao. We are grateful for their critical comments both in private correspondence and in published form.

As we expanded the scope of our initial paper it quickly became clear that we needed to conduct a number of interviews to fully flesh out our argument and ground it in current practices. We are indebted to Rod Alferness, David Auston, John Bowers, David Clarke, John English, Art Gossard, Herb Kroemer, James Langer,

Jim Merz, Matthew Tirrell, Pierre Wiltzius, and Michael Witherell of the University of California at Santa Barbara, for their willingness to share their experience of their institution.

At Janelia Research Campus Eric Betzig, Sean Eddy, Tim Harris, Ulrike Herberlein, Harald Hess, Tsung-Li Liu, Mark Phillip, Gerald Rubin, and Tanya Tabachnik all gave freely of their time and were very welcoming and open in their responses to our questions.

A number of Bell Labs alumni were very helpful in expanding our understanding of the Labs' singular culture. Some have already been acknowledged. In addition John Bean, Joe Campbell, Federico Capasso, Lloyd Harriot, Evelyn Hu, Debasis Mitra, and Julia Phillips provided invaluable insight into the culture of Bell Labs.

Ambuj Sagar and Laura Diaz Anadon undertook the selfless task of reviewing early versions of the entire manuscript. Marlee Chong was an amazing research assistant who contributed a great deal to this project. Patricia McLaughlin painstakingly went through every line and paragraph and Sarah Lefebvre kept things chugging along smoothly.

We are immensely grateful to Sheila Jasanoff for providing Tolu with an institutional home at the onset of his studies and for introducing us to one another. Our first editor at Harvard University Press, Michael Fisher, saw the possibility of this book in its earliest stages and provided encouragement. Thomas LeBien continued Michael's excellent work, provided a critical review of an early draft, and guided us through the sticky bits of book publishing at the press.

Finally, and most importantly, this book would have been impossible without the support of family. Onyi, Leri, and Leke all gave up their legitimate claims on Tolu's time and attention to allow this book to come into being; and for more than five decades Jaya has supported Venky's many career moves with ever present help and encouragement. Thanks to all of you. The credit for any positive contributions this book is able to make to the national conversation is rightfully yours.

INDEX

Quantized Electronic Structures
(QUEST), 108
quantum fluid, 52–55
quantum mechanics, 58–59

Rabi, Isidor, 61
radio astronomy, 45, 73
Radio Detection and Ranging (RADAR),
24–25
Randall, John, 24–25
RCA, 57, 71, 106
research: applied research, 1–3, 8–10,
20–38, 45–47, 67–70, 142–143; basic/
applied framework for, 12, 20–32; basic
research, 1–3, 8–10, 20–38, 45–47,
67–70, 142–143; bottlenecks in, 69;
boundaries in, 11–12, 14–19; categories for, 37–39; changing nomenclature
of, 34–35; classifying, 1–3, 8–10; culture of, 12, 56–57, 70–98; engineering
research, 14–19, 23–25, 48–49; explanation of, 11; funding, 4, 8–9, 44–45;
holistic view of, 10, 12, 14–15; linear
model and, 20–32, 44, 140, 143–144;
product development and, 11, 21–22,
46–47, 144–145; pure research, 37–38,
44–45; terms for, 17, 34–35; war and,
23–25. See also science
research environments: at Bell Labs,
10–13, 84–88; creating, 34; discovery-invention cycle and, 70–71; at
IBM, 60; importance of, 76, 134–140;
restructuring, 145–146; structure of,
137; sustaining, 76, 134–135
research institutions: building, 12, 99–130;
elements of, 12, 70–98, 105–111, 145;
innovative designs of, 99–129
research laboratories: elements of, 12,
70–98, 105–111, 145; first US laboratory, 15; Noble laureates and, 56–58;
role of, 37; success of, 124, 145
research time horizons, 68–69, 72
resonance method, 60–63
Restoring the Foundation (RtF), 140
Rising above the Gathering Storm, 131,
134–136
Rockwell International Science Center,
105
Rohrer, Heinrich, 59–60
Roosevelt, Franklin Delano, 3, 22–23
Rowland, Henry, 36, 41
Rubin, Gerald, 115–117, 123, 125

Rudenstein, Neil L., 17
Ruska, Ernst, 60

S&T policy, 3–4, 17, 20–21, 30–32. *See also*
Science and Technology policy
Sanger, Fred, 116
satellite technology, 73
scanning tunneling microscope (STM),
58–60
Schmitt, Roland, 59
Schumpeter, Joseph, 44
Schwartz, Lyle, 38–40
science: applied science, 1–3, 8–11, 20–47;
basic/applied framework for, 12,
20–32; basic science, 1–3, 8–11, 20–47;
boundaries in, 10–12, 20–32; holistic
view of, 12, 13; interconnectedness of,
48–49; pure science, 36, 41, 44. *See also*
research
Science and Technology Centers (STCs),
146
Science and Technology (S&T) policy:
action items for, 134–135; boundaries and, 17, 20–21; collaboration
and, 137–139; defining, 17; future of,
142–149; goals for, 137–140; multidisciplinary environments and, 136–139;
new framework for, 148–149; new
vision for, 30–32, 130–141, 144–149;
post-war policy, 3–4; reformulation
of, 130–141, 144–145; rethinking,
134–136, 144–145, 148–149; transdisciplinary environments and, 137–139
Science magazine, 9
Science, the Endless Frontier, 3, 10, 22–23,
25, 33–34, 36, 40, 44–46, 140
semiconductors, 49–51, 53–54, 57, 67
Shirakawa, Hideki, 67
Shockley, William, 2, 5–6, 8, 38, 49, 51–53,
55, 73–76
Siemens, 57
Smith, George, 14, 53, 55, 67, 98
Snow, C. P., 16
Solid State Lighting and Energy
Electronics Center (SSLEEC), 115
solid state physics, 58–59, 72, 131
Solow, Robert, 44
Spear, Ed, 103–104
steam engine, 27
Steinmetz, Charles, 37
Stern, Otto, 60–61
Stokes, Donald, 28–32, 40, 144